T0281359

Frontiers in Mathematics

K. David Elworthy
Yves Le Jan
Xue-Mei Li

The Geometry of Filtering

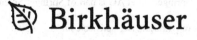 Birkhäuser

K. David Elworthy
Institute of Mathematics
University of Warwick
Gibbet Hill Road
CV4 7AL Coventry
United Kingdom
kde@maths.warwick.ac.uk

Yves Le Jan
Laboratoire de Mathématiques
Université Paris-Sud XI CNRS
Orsay Cedex
Bâtiment 425
France
yves.lejan@math.u-psud.fr

Xue-Mei Li
Institute of Mathematics
University of Warwick
Gibbet Hill Road
CV4 7AL Coventry
United Kingdom

Mathematics Subject Classification: 58J65, 60H07

ISBN 978-3-0346-0175-7   e-ISBN 978-3-0346-0176-4
DOI 10.1007/978-3-0346-0176-4

Cover design: deblik, Berlin

Printed on acid-free paper

Springer Basel AG is part of Springer Science+Business Media

www.birkhauser-science.com

Contents

Introduction

Filtering is the science of finding the law of a process given a partial observation of it. The main objects we study here are diffusion processes. These are naturally associated with second-order linear differential operators which are semi-elliptic and so introduce a possibly degenerate Riemannian structure on the state space. In fact, much of what we discuss is simply about two such operators intertwined by a smooth map, the "projection from the state space to the observations space", and does not involve any stochastic analysis.

From the point of view of stochastic processes, our purpose is to present and to study the underlying geometric structure which allows us to perform the filtering in a Markovian framework with the resulting conditional law being that of a Markov process which is time inhomogeneous in general. This geometry is determined by the symbol of the operator on the state space which projects to a symbol on the observation space. The projectible symbol induces a (possibly non-linear and partially defined) connection which lifts the observation process to the state space and gives a decomposition of the operator on the state space and of the noise. As is standard we can recover the classical filtering theory in which the observations are not usually Markovian by application of the Girsanov-Maruyama-Cameron-Martin Theorem.

This structure we have is examined in relation to a number of geometrical topics. In one direction this leads to a generalisation of Hermann's theorem on the fibre bundle structure of certain Riemannian submersions. In another it gives a novel description of generalised Weitzenböck curvature. It also applies to infinite dimensional state spaces such as arise naturally for stochastic flows of diffeomorphisms defined by stochastic differential equations, and for certain stochastic partial differential equations.

A feature of our approach is that in general we use canonical processes as solutions of martingale problems to describe our processes, rather than stochastic differential equations and semi-martingale calculus, unless we are explicitly dealing with the latter. This leads to some new constructions, for example of integrals along the paths of our diffusions in Section 4.1, which are valid more generally than in the very regular cases we discuss here.

Those whose interest is mainly in filtering rather than in the geometry should look at Chapter 1, most of Chapter 2, especially Section 2.3, but omitting Section 2.6. Then move to Chapter 4 where Section 4.11 can be ignored. They could then finish with Chapter 5, though some of the Appendices may be of interest.

A central role is played by certain generalised connections determined by the principal symbols of the operators involved. To describe this in more detail let M be a smooth manifold. Consider a smooth second-order semi-elliptic differential operator $\mathcal{L}$ such that $\mathcal{L}1 \equiv 0$. In a local chart, such an operator takes the form

$$\mathcal{L} = \frac{1}{2} \sum_{i,j=1}^{n} a^{ij} \frac{\partial}{\partial x^i} \frac{\partial}{\partial x^j} + \sum b^i \frac{\partial}{\partial x^i} \tag{1}$$

where the a^{ij}'s and b^i's are smooth functions and the matrix (a^{ij}) is positive semi-definite.

Such differential operators are called diffusion operators. An elliptic diffusion operator induces a Riemannian metric on M. In the degenerate case we shall have to assume that the "symbol" of $\mathcal{L}$ (essentially the matrix $[a^{ij}]$ in the representation (1)) has constant rank and so determines a subbundle E of the tangent bundle TM together with a Riemannian metric on E. In Elworthy-LeJan-Li [35] and [36] it was shown that a diffusion operator in Hörmander form, satisfying this condition, induces a linear connection on E which is adapted to the Riemannian metric induced on E, but not necessarily torsion free. It was also shown that all metric connections on E can be constructed by some choice of Hörmander form for a given $\mathcal{L}$ in this way. The use of such connections has turned out to be instrumental in the decomposition of noise and calculation of covariant derivatives of the derivative flows.

A related construction of connections can arise with principal fibre bundles P. An equivariant differential operator on P induces naturally a diffusion operator on the base manifold. Conversely, given an equivariant or "principal" connection on P, one can lift horizontally a diffusion operator on the base manifold of the form of sum of squares of vector fields by simply lifting up the vector fields. It still needs to be shown that the lift is independent of choices of its Hörmander form. Consider now a diffusion operator not given in Hörmander form. Since it has no zero-order term we can associate with it an operator δ which sends differential one-forms to functions. In Proposition 1.2.1, members of a class of such operators are described, each of which determines a diffusion operator. Horizontal lifts of diffusion operators can then be defined in terms of the δ operator. This construction extends to situations where there is no equivariance and we have only partially defined and non-linear connections. We show that given a smooth $p : N \to M$: a diffusion operator $\mathcal{B}$ on N which lies over a diffusion operator $\mathcal{A}$ on M satisfying a "cohesiveness" property gives rise to a *semi-connection*, a partially defined, non-linear, connection which can be characterised by the property that, with respect to it, $\mathcal{B}$ can be written as the direct sum of the horizontal lift of its induced operator and a vertical diffusion operator. Of particular importance are examples where $p : N \to M$ is a principal bundle. In that case the vertical component of $\mathcal{B}$ induces differential operators on spaces of sections of associated vector bundles: we observe that these are zero-order operators, and can have geometric significance.

This geometric significance, and the relationship between these partially defined connections and the metric connections determined by the Hörmander form as in [35] and [36], is seen when taking $\mathcal{B}$ to be the generator of the diffusion given on the frame bundle GLM of M by the action of the derivative flow of a stochastic differential equation on M. The semi-connection determined by $\mathcal{B}$ is then equivariant and is the *adjoint* of the metric connection induced by the SDE in a sense extending that of Driver [25] and described in [36]. The zero-order operators induced by the vertical component of $\mathcal{B}$ acting on differential forms turn out to be generalised Weitzenböck curvature operators, in the sense of [36], reducing to the classical ones when M is Riemannian for particular choices of stochastic differential equations for Brownian motion on M. Our filtering then reproduces the conditioning results for derivatives of stochastic flows in [38] and [36].

Our approach is also applied to the case where M is compact and N is its diffeomorphism group, $\mathrm{Diff}(M)$, with P evaluation at a chosen point of M. The operator $\mathcal{B}$ is taken to be the generator of the diffusion process on $\mathrm{Diff}(M)$ arising from a stochastic flow. However our constructions can be made in terms of the reproducing Hilbert space of vector fields on M defined by the flow. From this we see that stochastic flows are essentially determined by a class of semi-connections on the bundle $p : \mathrm{Diff}(M) \to M$ and smooth stochastic flows whose one-point motions have a cohesive generator determine semi-connections on all natural bundles over M. Apart from these geometrical aspects of stochastic flows we also obtain a skew-product decomposition which, for example, can be used to find conditional expectations of functionals of such flows given knowledge of the one-point motion from our chosen point in M.

The plan of the book is as follows: In Chapter 1 we describe various representations of diffusion operators and when they are available. We also define the notion of such an operator being *along a distribution*. In Chapter 2 we introduce the notion of *semi-connection* which is fundamental for what follows, and we show how these are induced by certain intertwined pairs of diffusion operators and how they relate to a canonical decomposition of such operators. We also have a first look at the topological consequences on $p : N \to M$ of having $\mathcal{B}$ on N over some $\mathcal{A}$ on M which possesses hypo-ellipticity type properties. This is a minor extension of part of Hermann's theorem, [51], for Riemannian submersions. In Chapter 3 we specialise to the case of principal bundles, introduce the example of derivative flow, and show how the generalised Wietzenbock curvatures arise.

It is not really until Chapter 4 that stochastic analysis plays a major role. Here we describe methods of conditioning functionals of the $\mathcal{B}$-process given information about its projection onto M. We also use our decomposition of $\mathcal{B}$ and resulting decomposition of the $\mathcal{B}$-process to describe the conditional $\mathcal{B}$-process.

In the equivariant case of principal bundles the decomposition of the process can be considered as a skew-product decomposition. In Chapter 5 we show how our constructions can apply to classical filtering problems, where the projection of the $\mathcal{B}$-process is non-Markovian by an appropriate change of probability measure. We can follow the classical approach, illustrated for example in the lecture notes of Pardoux [85], and obtain, in Theorem 5.9, a version of Kushner's formula for non-linear filtering in somewhat greater generality than is standard. This requires some discussion of analogues of *innovations* processes in our setting.

We return to more geometrical analysis in Chapter 6, giving further extensions of Hermann's theorem and analysing the consequences of the horizontal lift of A commuting with B, thereby extending the discussion in [9]. In particular we see that such commutativity, plus hypo-ellipticity conditions on $\mathcal{A}$, gives a bundle structure and a diffusion operator on the fibre which is preserved by the trivialisations of the bundle structure. This leads to an extension of the "skew-product" decomposition given in [33] for Brownian motions on the total space of Riemannian submersions with totally geodesic fibres. In fact the well-known theory for Riemann submersions, and the special case arising from Riemannian symmetric spaces is presented in Chapter 7.

Chapter 8 is where we describe the theory for the diffeomorphism bundle $p : \text{Diff}(M) \to M$ with a stochastic flow of diffeomorphism on M. Initially this is done independently of stochastic analysis and in terms of reproducing kernel Hilbert spaces of vector fields on M. The correspondence between such Hilbert spaces and stochastic flows is then used to get results for flows and in particular skew-product decompositions of them.

In the Appendices we present the Girsanov Theorem in a way which does not rely on having to use conditions such as Novikov's criteria for it to remain valid. This has been known for a long time, but does not appear to be as well known as it deserves. We also look at conditions for degenerate, but smooth, diffusion operators to have smooth Hörmander forms, and so to have stochastic differential equation representations for their associated processes. We also discuss semi-martingales and Γ-martingales along a subbundle of the tangent bundle with a connection. One section of the Appendix is a very brief exposition of the differential geometry of submanifolds, defining second fundamental forms and shape operators. This is used in the final section which analyses the situation of intertwined stochastic flows or essentially equivalently of diffusion operators which are not only intertwined but also have Hörmander forms composed of intertwined vector fields. It is shown that the Hörmander forms determine a decomposition of the operator $\mathcal{B}$, which is not generally the same as the canonical decomposition described in Chapter 2. Here it is not necessary to make the constant rank condition on the symbol of $\mathcal{A}$ which plays an important role in Chapter 2. At the end of this section we show that having intertwined Brownian flows which both induce Levi-Civita connections can only occur given severe restrictions on the geometry of the submersion $p : N \to M$.

For Brownian motions on the total spaces of Riemannian submersions much of our basic discussion, as in the first two and a half chapters, of skew-product decompositions is very close to that in [33] which was taken further by Liao in [69]. A major difference from Liao's work is that for degenerate diffusions we use the semi-connection determined by our operators rather than an arbitrary one, so obtaining canonical decompositions. The same holds for the very recent work of Lazaro-Cami & Ortega, [62] where they are motivated by the reduction and reconstruction of Hamiltonian systems and consider similar decompositions for semi-martingales. An extension of [33] in a different direction, to shed light on the Fadeev-Popov procedure for gauge theories in theoretical physics, was given by Arnaudon &Paycha in [1]. Much of the equivariant theory presented here was announced with some sketched proofs in [34].

Acknowledgements

The collaborative work for this book was carried out at a variety of institutions, especially Orsay and Warwick, with final touches at the Newton Institute. We gratefully acknowledge the importance of our contacts during its preparation with many people, notably including J.-M.Bony, D.O.Crisan, R.Dalang, M.Fuhrman, E.P. Hsu, and J.Teichmann. This research also benefited from an EPSRC grant (EP/E058124/1).

K.D. Elworthy, Y. Le Jan and Xue-Mei Li

Chapter 1

Diffusion Operators

Let $\sum_{i,j=1}^{n} a^{ij} \frac{\partial}{\partial x^i} \frac{\partial}{\partial x^j} + \sum_{i=1}^{n} b^i \frac{\partial}{\partial x^i}$ be a differential operator on $\mathbf{R}^n$, with smooth coefficients. We can and will assume that $(a^{i,j})$ is symmetric. Its *symbol* is given by the matrix-valued function $(a^{i,j})$ considered as a bilinear form on $\mathbf{R}^n$ or equivalently as a map from $(\mathbf{R}^n)^\star$ to $\mathbf{R}^n$. The operator is said to be *semi-elliptic* if the symbol is positive semi-definite, and *elliptic* if it is positive definite. More generally if $\mathcal{L}$ is a second-order differential operator on a manifold M, denote by $\sigma^{\mathcal{L}} : T^*M \to TM$ its symbol determined by

$$df\left(\sigma^{\mathcal{L}}(dg)\right) = \frac{1}{2}\mathcal{L}(fg) - \frac{1}{2}(\mathcal{L}f)g - \frac{1}{2}f(\mathcal{L}g),$$

for C^2 functions f, g. We will often write $\sigma^{\mathcal{L}}(\ell^1, \ell^2)$ for $\ell^1 \sigma^{\mathcal{L}}(\ell^2)$ and consider $\sigma^{\mathcal{L}}$ as a bilinear form on T^*M. Note that it is symmetric. The operator is said to be semi-elliptic if $\sigma^{\mathcal{L}}(\ell^1, \ell^2) \geqslant 0$ for all $\ell^1, \ell^2 \in T_u M^*$, all $u \in M$, and elliptic if the inequality holds strictly. Ellipticity is equivalent to $\sigma^{\mathcal{L}}$ being onto.

Definition 1.0.1. A semi-elliptic smooth second-order differential operator $\mathcal{L}$ is said to be a *diffusion operator* if $\mathcal{L}1 = 0$.

The standard example of a diffusion operator is the *Laplace-Beltrami operator* or *Laplacian*, $\triangle$, of a Riemannian manifold M. It is given as in $\mathbf{R}^n$ by $\triangle = \mathrm{div}\,\nabla = -d^*d$ where div is the negative of the adjoint of the gradient operator ∇. Its symbol $\sigma^\triangle : T^*M \to TM$ is just the isomorphism induced by the Riemannian metric. Conversely the symbol of any elliptic diffusion operator determines a Riemannian metric with respect to which the operator differs from the Laplacian by a first-order term.

1.1 Representations of Diffusion Operators

Apart from local representations as given above there are several global ways to represent a diffusion operator $\mathcal{L}$. One is to take a connection ∇ on TM. Recall that

K.D. Elworthy et al., *The Geometry of Filtering*, Frontiers in Mathematics,
DOI 10.1007/978-3-0346-0176-4_1, © Springer Basel AG 2010

a *connection* on TM gives, or is given by, a *covariant derivative operator* ∇ acting on vector fields. For each C^r vector field U on M it gives a C^{r-1} section $\nabla_- U$ of $\mathbb{L}(TM; TM)$. In other words, for each $x \in M$ we have a linear map $v \mapsto \nabla_v U$ of $T_x M$ to itself. This covariant derivative of U in the direction v satisfies the usual rules. In particular it is a derivation with respect to multiplication by differentiable functions $f : M \to \mathbf{R}$, so that $\nabla_v(fU) = df(v)U(x) + f(x)\nabla_v U$.

For example on $\mathbf{R}^n$ a connection is given by Christoffel symbols $\Gamma_{ij}^k : \mathbf{R}^n \to \mathbf{R}$, $i, j, k = 1, \ldots, n$. These define a covariant differentiation of vector fields by

$$(\nabla_v U)^k = \sum_{i=1}^n \left(\frac{\partial U^k}{\partial x_i}(x) + \sum_{j=1}^n \Gamma_{ij}^k(x) U^j(x) \right) v^i \tag{1.1}$$

where, for example, U^j denotes the j-th component of the vector field U and we are considering v as a tangent vector at $x \in \mathbf{R}^n$. We can also consider the Christoffel symbols as the components of a map $\Gamma : \mathbf{R}^n \to \mathbb{L}(\mathbf{R}^n; \mathbb{L}(\mathbf{R}^n; \mathbf{R}^n))$ and equation (1.1) becomes:

$$\nabla_v U = DU(x)(v) + \Gamma(x)(v)(U(x)). \tag{1.2}$$

Given any smooth vector bundle $\tau : E \to M$ over M a *connection on E* gives a similar covariant derivative acting on sections U of E. This time $v \mapsto \nabla_v U$ is in $\mathbb{L}(T_x M; E_x)$, where E_x is the fibre over x for $x \in M$. In our local representation the Christoffel symbol now has $\Gamma(x) \in \mathbb{L}(\mathbf{R}^n; \mathbb{L}(E_x; E_x))$ where $E_x := p^{-1}(x)$ is the fibre of E over x. A connection on TM determines one on the cotangent bundle $T^* M$, and on all tensor bundles over M.

Such connections always exist; more details can be found in [19], [55].

In terms of a connection on TM we can write

$$\mathcal{L}f(x) = \text{trace}_{T_x M}\nabla_-(\sigma^{\mathcal{L}}(df)) + df(V^0(x)) \tag{1.3}$$

for some smooth vector field V^0 on M. The trace is that of the mapping $v \mapsto \nabla_v(\sigma^{\mathcal{L}}(df))$ from $T_x M$ to itself. To see this it is only necessary to check that the right-hand side has the correct symbol since the symbol determines the diffusion operator up to a first-order term. For the Laplacian on $\mathbf{R}^n$, when $\mathbf{R}^n$ is given its usual "trivial" connection this reduces to the representation:

$$\triangle f(x) = \text{trace}\, D^2 f(x) \tag{1.4}$$

where $D^2 f(x) \in \mathbb{L}(\mathbf{R}^n, \mathbf{R}^n; \mathbf{R})$ is the second Frechet derivative of f at x.

If a smooth 'square root' to $2\sigma^{\mathcal{L}}$ can be found we have a Hörmander representation. The 'square root' is a smooth map $X : M \times \mathbf{R}^m \to TM$ with each $X(x) \equiv X(x, -) : \mathbf{R}^m \to T_x M$ linear, such that

$$2\sigma_x^{\mathcal{L}} = X(x)X(x)^* : T_x^* M \to T_x M$$

where $X(x)^* : T_x^* M \to \mathbf{R}^m$ is the adjoint map of $X(x)$. Thus there is a smooth vector field A with

$$\mathcal{L} = \frac{1}{2} \sum_{j=1}^{m} \mathbf{L}_{X^j} \mathbf{L}_{X^j} + \mathbf{L}_A, \tag{1.5}$$

where $X^j(x) = X(x)(e_j)$ for $\{e_j\}$ an orthonormal basis of $\mathbf{R}^m$, and $\mathbf{L}_V$ denotes *Lie differentiation* with respect to a vector field V, so $\mathbf{L}_V f(x) = df_x(V(x))$. On $\mathbf{R}^n$,

$$\mathbf{L}_V f(x) = \sum_i V^i(x) \frac{\partial}{\partial x_i} f(x).$$

For the Laplacian on $\mathbf{R}^n$ the simplest Hörmander form representation is

$$\triangle = \frac{1}{2} \sum_{j=1}^{n} \sqrt{2} \frac{\partial}{\partial x^j} \sqrt{2} \frac{\partial}{\partial x^j}$$

and the generator, $\frac{1}{2}\triangle$, of Brownian motion has

$$\frac{1}{2}\triangle = \frac{1}{2} \sum_{j=1}^{n} \frac{\partial}{\partial x^j} \frac{\partial}{\partial x^j}.$$

If $\sigma^{\mathcal{L}}$ has constant rank, such X may be found. A proof of this is given in the Appendix, Theorem 9.2.1. Otherwise it is only known that locally Lipschitz square roots exist (see the discussions in the Appendix, Section 9.2). In that case $\mathbf{L}_{X^j}\mathbf{L}_{X^j}$ is only defined almost surely everywhere and the vector field A can only be assumed measurable and locally bounded. Nevertheless uniqueness of the martingale problem still holds (see below). Also there is still the hybrid representation, given a connection ∇ on TM,

$$\mathcal{L}f(x) = \frac{1}{2} \sum_{j=1}^{m} \nabla_{X^j(x)}(df)(X^j(x)) + df(V^0(x)) \tag{1.6}$$

for V^0 locally Lipschitz.

The choice of a Hörmander representation for a diffusion operator, if it exists, determines a locally defined stochastic flow of diffeomorphisms $\{\xi_t : 0 \leqslant t < \zeta\}$ whose one-point motion solves the martingale problem for the diffusion operator. In particular on bounded measurable compactly supported $f : M \to \mathbf{R}$ the associated (sub-) Markovian semigroup is given by $P_t f = \mathbf{E}(f \circ \xi_t)$. See also Appendix II.

Despite the discussion above we can always write $\mathcal{L}$ in the form

$$\mathcal{L} = \frac{1}{2} \sum_{ij=1}^{N} a^{ij} \mathbf{L}_{X^i} \mathbf{L}_{X^j} + \mathbf{L}_{X^0}, \tag{1.7}$$

where N is a finite number, a^{ij} and X^k are respectively smooth functions and smooth vector fields with $a^{ij} = a^{ji}$.

1.2   The Associated First-Order Operator

Denote by $C^r \Lambda^p \equiv C^r \Lambda^p T^* M$, $r \geqslant 0$, the space of C^r smooth differential p-forms on a manifold N. To each diffusion operator $\mathcal{L}$ we shall associate an operator $\delta^{\mathcal{L}} : C^{r+1} \Lambda^1 \to C^r(M; \mathbf{R})$, see Elworthy-LeJan-Li [35], [36] c.f. Eberle [27]. The horizontal lift of $\mathcal{L}$ will then be defined in terms of a lift of $\delta^{\mathcal{L}}$. The existence of $\delta^{\mathcal{L}}$ comes from the lack of a 0-order term in $\mathcal{L}$:

Proposition 1.2.1. *For each diffusion operator $\mathcal{L}$ there is a unique smooth linear differential operator $\delta^{\mathcal{L}} : C^{r+1} \Lambda^1 \to C^r \Lambda^0$ such that*

(1) $\delta^{\mathcal{L}} (f\phi) = df \sigma^{\mathcal{L}}(\phi) + f \cdot \delta^{\mathcal{L}} (\phi)$,

(2) $\delta^{\mathcal{L}} (df) = \mathcal{L}f$.

In particular in the $\mathbf{R}^n$ case with $\mathcal{L} = \sum_{i,j=1}^n a^{ij} \frac{\partial}{\partial x^i} \frac{\partial}{\partial x^j} + \sum b^i \frac{\partial}{\partial x^i}$ it is given by

$$\delta^{\mathcal{L}} \phi = \sum_{i,j=1}^m a^{ij} \frac{\partial}{\partial x_i} \phi_j(x) + \sum b^i \phi_i(x)$$

where ϕ has the representation

$$\phi_x = \sum \phi_j(x) \, dx^i. \tag{1.8}$$

Equivalently $\delta^{\mathcal{L}}$ is determined by either one of the following:

$$\delta^{\mathcal{L}}(fdg) = \sigma^{\mathcal{L}}(df, dg) + f\mathcal{L}g, \tag{1.9}$$

$$\delta^{\mathcal{L}}(fdg) = \frac{1}{2}\mathcal{L}(fg) - \frac{1}{2}g\mathcal{L}f + \frac{1}{2}f\mathcal{L}g. \tag{1.10}$$

Proof. The statements are rather obvious in the R^n case. In general take a connection ∇ on TM, then, as in (1.3), $\mathcal{L}$ can be written as $\mathcal{L}f = \text{trace} \nabla \sigma^{\mathcal{L}}(df) + \mathbf{L}_{V^0} f$ for some smooth vector field V^0. Set

$$\delta^{\mathcal{L}} \phi = \text{trace} \nabla (\sigma^{\mathcal{L}} \phi) + \phi(V^0).$$

Then $\delta^{\mathcal{L}}(df) = \mathcal{L}f$ and

$$\delta^{\mathcal{L}}(f\phi) = \text{trace} \nabla (f(\sigma^{\mathcal{L}}\phi)) + f\phi(V^0) = f\delta^{\mathcal{L}}\phi + df(\sigma^{\mathcal{L}}\phi).$$

Note that a general C^r 1-form ϕ can be written as $\phi = \sum_{j=1}^k f_i dg_i$ for some C^r function f_i and smooth g_i, for example, by taking $(g^1, \ldots, g^m) : M \to \mathbf{R}^m$ to be an immersion. This shows that (1) and (2) determine $\delta^{\mathcal{L}}$ uniquely. Moreover since $\mathcal{L}$ is a smooth operator so is $\delta^{\mathcal{L}}$. □

Remark 1.2.2. If the diffusion operator $\mathcal{L}$ has a representation

$$\mathcal{L} = \sum_{j=1}^m a^{ij} \mathbf{L}_{X^i} \mathbf{L}_{X^j} + \mathbf{L}_{X^0}$$

for some smooth vector fields X^i and smooth functions a^{ij}, $i, j = 0, 1, \ldots, m$ with $a^{ij} = a^{ji}$, then

$$\delta^{\mathcal{L}} = \sum_{j=1}^{m} a^{ij} \mathbf{L}_{X^i} \iota_{X^j} + \iota_{X^0},$$

where ι_A denotes the interior product of the vector field A with a differential form. [In particular if ϕ is a one-form, then $\iota_A(\phi) : M \to \mathbf{R}$ is given by $\iota_A(\phi)(x) = \phi_x(A(x))$.] One can check directly that $\delta^{\mathcal{L}}(df) = \mathcal{L}f$ and that (1) holds.

1.3 Diffusion Operators Along a Distribution

Let N be a smooth manifold. By a **distribution** S in N we mean a family $\{S_u : u \in N\}$ where S_u is a linear subspace of $T_u N$; for example S could be a subbundle of TN. In $\mathbf{R}^n$ each S_u can be viewed as a linear subspace of R^n. Given such a distribution S let $S^0 = \cup_u S_u^0$ for S_u^0 **the annihilator** of S_u in $T_u^* N$.

Definition 1.3.1. Let S be a distribution in TN. Denote by $C^r S^0$ the set of C^r 1-forms which vanish on S. A diffusion operator $\mathcal{L}$ on N is said to be **along** S if $\delta^{\mathcal{L}} \phi = 0$ for all $\phi \in C^1 S^0$.

Example 1.3.2. For $N = \mathbf{R}^n - \{0\}$ let S_u be the hyperplane orthogonal to u. The spherical Laplacian, $(\frac{\partial^2}{\partial \theta^2}$ if $n = 2)$, is along this distribution.

Example 1.3.3. Let $N = \mathbf{R}^3$ with the C^∞ *Heisenberg group* structure induced by the central extension of the symplectic vector space $\mathbf{R}^2$. This is defined by

$$(x, y, z) \cdot (x', y', z') = \left(x + x', y + y', z + z' + \frac{1}{2}(xy' - yx')\right).$$

This is isomorphic to the matrix group of 3×3 upper diagonal matrices:

$$(x, y, z) \mapsto \begin{pmatrix} 1 & x & z \\ 0 & 1 & y \\ 0 & 0 & 1 \end{pmatrix}. \tag{1.11}$$

Let X, Y, Z be the left-invariant vector fields which give the standard basis for $\mathbf{R}^3$ at the origin. As operators:

$$X(x, y, z) = \frac{\partial}{\partial x} - \frac{1}{2} y \frac{\partial}{\partial z}, \qquad Y(x, y, z) = \frac{\partial}{\partial y} + \frac{1}{2} x \frac{\partial}{\partial z}$$

$$Z(x, y, z) = \frac{\partial}{\partial z}.$$

Let $\mathcal{L}$ be the operator $\mathcal{L}^{\mathbb{H}}$ given by

$$\mathcal{L}^{\mathbb{H}} = \frac{1}{2} \left(\mathbf{L}_X \mathbf{L}_X + \mathbf{L}_Y \mathbf{L}_Y \right) \tag{1.12}$$

$$= \frac{1}{2} \left(\frac{\partial^2}{\partial x^2} + \frac{\partial^2}{\partial y^2} + \frac{1}{4}(x^2 + y^2) \frac{\partial^2}{\partial z^2} + x \frac{\partial^2}{\partial y \partial z} - y \frac{\partial^2}{\partial x \partial z} \right). \tag{1.13}$$

This operator is clearly along the distribution S given by left translates of the (x, y)-plane. This distribution is not integrable: it is not tangent to any foliation of $\mathbf{R}^3$. Indeed the Lie bracket of X and Y is Z so $[X, Y](u) \notin S_u$.

In general suppose $\mathcal{L}$ is along S and take $\phi \in C^r S^0$. By Proposition 1.2.1 and the symmetry of $\sigma^{\mathcal{L}}$, $0 = (df)(\sigma^{\mathcal{L}}(\phi)) = \phi(\sigma^{\mathcal{L}}(df)$ giving $\phi_x \in \text{Image}[\sigma_x^{\mathcal{L}}]^0$. This proves Remark 1.3.4 (i):

Remark 1.3.4.   (i) if $\delta^{\mathcal{L}}\phi = 0$ for all $\phi \in C^1 S^0$, then $\sigma^{\mathcal{L}}\phi = 0$ for all such ϕ and $\text{Image}[\sigma_x^{\mathcal{L}}] \subset \cap_{\phi \in C^1 S^0}[\ker \phi_x]$ for all $x \in N$.

(ii) If S is a subbundle of TN, (essentially a smooth family of subspaces of constant dimension), and $\mathcal{L}$ is along S, then without ambiguity we can define $\delta^{\mathcal{L}}\phi$ for ϕ a C^0 section of S^* by $\delta^{\mathcal{L}}\phi := \delta^{\mathcal{L}}\tilde{\phi}$ for any 1-form $\tilde{\phi}$ extending ϕ. Recall that S^* is canonically isomorphic to the quotient T^*N/S^0. An important case is where S_u is the image of $\sigma_u^{\mathcal{L}}$ assuming that the symbol $\sigma^{\mathcal{L}}$ has constant rank.

Definition 1.3.5. If
$$S_x = \cap_{\phi \in C^1 S^0}[\ker \phi_x]$$
for all x we say S is a **regular distribution**.

Clearly subbundles are regular. As another example take the circle, $M = S^1$, and suppose that $S_u = 0$ except for finitely many points. Then $C^1 S^0$ consists of those C^1 forms which vanish at those points. This distribution is regular but is not a subbundle. However this would not hold in general if it were non-zero precisely on a countably infinite set of points. The notion is introduced in order to be able to consider the **vertical distribution** $\{VT_u N := \ker T_u p : u \in N\}$ of a smooth map $p : N \to M$:

Lemma 1.3.6. *Let $p : N \to M$ be a smooth map, then $\{\ker T_u p : u \in N\}$ is a regular distribution.*

Proof. This is immediate since $\ker[Tp]$ is annihilated by all differential 1-forms of the form $\theta \circ Tp$ for θ a C^1 1-form on M. $\square$

Proposition 1.3.7.   (1) *Let S be a regular distribution of N and $\mathcal{L}$ an operator written in Hörmander form:*

$$\mathcal{L} = \frac{1}{2} \sum_{j=1}^{m} \mathbf{L}_{Y^j} \mathbf{L}_{Y^j} + \mathbf{L}_{Y^0} \tag{1.14}$$

where the vector fields Y^0 and $Y^j, j = 1, \ldots, m$ are C^0 and C^1 respectively. Then $\mathcal{L}$ is along S if and only if Y^i are sections of S.

(2) *If $\mathcal{B}$ is along a smooth subbundle S of TN, then for any connection ∇^S on S we can write $\mathcal{B}$ as*

$$\mathcal{B}f = \text{trace}_{S_x} \nabla^S_{-}\left(\sigma^{\mathcal{B}}(df)\right) + \mathbf{L}_{X^0} f.$$

Also we can find smooth sections $X^0, \ldots, X^m$ of S and smooth functions a^{ij} such that

$$\mathcal{B} = \frac{1}{2} \sum_{i,j} a^{ij} \, \mathbf{L}_{X^i} \mathbf{L}_{X^j} + \mathbf{L}_{X^0}.$$

[Recall that ∇^S is a covariant derivative operator defined only on those vector fields which take values in S, though one can differentiate in any direction.]

Proof. For part (1), if Y^i are sections of S, take $\phi \in C^1 S^0$, then

$$\delta^{\mathcal{L}} \phi = \frac{1}{2} \sum_{j=1}^{m} \mathbf{L}_{Y^j} \phi(Y^j) + \phi(Y^0) = 0$$

and so $\mathcal{L}$ is along S.

Conversely suppose $\mathcal{L}$ is along S. Define a C^1 bundle map $Y : \mathbf{R}^m \to TN$ by $Y(x)(e) = \sum_{j=1}^{m} Y^j(x) e_j$ for $\{e_j\}_{j=1}^{m}$ an orthonormal base of $\mathbf{R}^m$. Then

$$2\sigma_x^{\mathcal{L}} = Y(x) Y(x)^*$$

and

$$\text{Image}[Y(x)] = \text{Image}[\sigma_x^{\mathcal{L}}] \subset S,$$

by Remark 1.3.4. Now

$$\delta^{\mathcal{L}} \phi = \frac{1}{2} \sum \mathbf{L}_{Y^j}(\phi(Y^j)) + \phi(Y^0) = \phi(Y^0),$$

which can only vanish for all $\phi \in C^1 S^0$ if Y^0 is a section of S. Thus $Y^1, \ldots, Y^m$, and Y^0 are all sections of S.

For part (2), we use (1.3) and take ∇ there to be the direct sum of ∇^S with an arbitrary connection on a complementary bundle, observing $\sigma^{\mathcal{B}}$ has image in S by Remark 1.3.4(i). □

1.4 Lifts of Diffusion Operators

Let $p : N \to M$ be a smooth map and E a subbundle of TM. Recall that $VT_u N := \ker T_u p$ denotes the vertical distribution. If $p(u)$ is a *regular value* of p, that is if $T_y p : T_y N \to T_{p(u)} M$ is surjective for all $y \in p^{-1}(p(u))$, then the fibre $E_p(u) := p^{-1}(p(u))$ is a submanifold with $VT_u N$ as its tangent space at u. Let S be a subbundle of TN transversal to the fibre of p, i.e. $VT_u N \cap S = \{0\}$ all $u \in N$ and such that $T_y p$ maps S_y isomorphically onto $E_{p(y)}$, for each y.

Example 1.4.1. Set $N = \mathbf{R}^2 - \{0\}$, $M = \mathbf{R}(> 0)$ with $E_x = \mathbf{R}$ for all $x \in M$. Take $p(x) = |x|$. Then for all $u \in N$ the orthogonal complement to $\mathbf{R}u$ is the vertical tangent spaced $VT_u N$ and we may take $S_u = \mathbf{R}u$. Here we are, as usual, identifying $T_u\{\mathbf{R}^2 - \{0\}\}$ with $\mathbf{R}^2$.

The example above was a special case of the situation when we have a Riemannian metric on N, and so an inner product, $\langle -, - \rangle_u$, on each tangent space $T_u N$ and when $E = TM$. If, as in the example, p is a *submersion* , i.e. its derivative $T_u p : T_u N \to T_{p(u)} M$ is onto for each u, we can take each S_u to be the orthogonal complement in T_u of $VT_u N$. If p were not a submersion, no transversal subbundle S would exist. Such submersions are described in detail in Chapter 7 with examples. Here is another one:

Example 1.4.2. Let $N = SO(3) - \{\text{Id}\}$, be the special orthogonal group of $\mathbf{R}^3$ with identity element removed. Let $M = \mathbf{RP}^2$, real 2-dimensional projective space, considered as the space of lines through the origin in $\mathbf{R}^3$ or equivalently as the 2-sphere with antipodal points identified. Define $p : N \to \mathbf{RP}^2$ by taking $p(u)$ to be the axis of rotation of u, i.e. the line determined by the eigenvector with eigenvalue 1. The fibre through any $u \in N$ is a copy of the rotation group $SO(2)$, that is, of S^1, with the identity removed. In fact it is part of a one-parameter subgroup $\{e^{t\alpha} : t \in \mathbf{R}\}$, say, of $SO(3)$ through u. Here α is an element of the Lie algebra $\mathfrak{so}(3)$. We can take S_u to be the left translate of the orthogonal complement of α in $\mathfrak{so}(3)$, identified with $T_{Id} SO(3)$, to u. It is identified under Tp with the plane in $\mathbf{R}^3$ orthogonal to $p(u)$, which is naturally identified with $T_{p(u)} \mathbf{RP}^2$.

For another class of examples see Example 2.1.6 below.

Lemma 1.4.3. *Every smooth 1-form on N can be written as a linear combination of sections of the form $\psi + \lambda p^*(\phi)$ for $\lambda : N \to \mathbf{R}$ smooth, ϕ a 1-form on M, and ψ annihilates S. In particular any 1-form annihilating VTN is of the form $\lambda p^*(\phi)$. If $E = TM$, then ψ is uniquely determined.*

Proof. Take Riemannian metrics on M and N such that the isomorphism between S and $p^*(E)$ given by Tp is isometric. Fix $y_0 \in N$. Take a neighbourhood V of $p(y_0)$ in M over which E is trivializable. Let $v^1, v^2, \ldots, v^p$ be a trivialising family of sections over V. Set $U = p^{-1}(V)$. If $\phi^j = (v^j)^*$, the dual 1-form to v^j, $j = 1$ to p, over V, then $\{p^*(\phi^j)^\#, j = 1 \text{ to } p\}$ gives a trivialization of S over U. [Indeed $p^*(\phi^j)_y(-) = \phi^j_{p(y)}(T_y p-) = \langle (T_y p)^*(v^j), - \rangle$.] Since any vector field over V can therefore be written as one orthogonal to S plus a linear combination of the $p^*(\phi^j)^\#$, by duality the result holds for forms with support in U. The global result follows using a partition of unity.

For the uniqueness note that if $E = TM$, then $TN = VTN + S$. $\square$

By a **lift** of a diffusion operator $\mathcal{A}$ on M over p we mean a diffusion operator $\mathcal{B}$ on N such that

$$\mathcal{B}(f \circ p) = (\mathcal{A}f) \circ p \tag{1.15}$$

for all C^2 functions f on M.

Proposition 1.4.4. *Let $\mathcal{A}$ be a diffusion operator on M along the subbundle E of TM. There is a unique lift of $\mathcal{A}$ to a smooth diffusion generator $\mathcal{A}^S$ along the transversal bundle S. Write $\bar{\delta} = \delta^{\mathcal{A}^S}$. Then $\mathcal{A}^S$ is determined by*

(i) $\bar{\delta}(\psi) = 0$ *if* ψ *annihilates* S.

(ii) $\bar{\delta}(p^*\phi) = (\delta^A \phi) \circ p,$ *for* $\phi \in \Omega^1(M)$.

Moreover

(iii) *for* $y \in N$ *let* $h_y : E_{p(y)} \to T_y N$ *be the right inverse of* $T_y p$ *with image* S_y. *Then*

    (a) $\sigma_y^{A^S} = h_y \, \sigma^A \, h_y^*$.

    (b) *If* A *is given by*

$$A = \frac{1}{2} \sum_{i,j=1}^{N} a^{ij} \mathbf{L}_{X^i} \mathbf{L}_{X^j} + \mathbf{L}_{X^0} \tag{1.16}$$

    where $X^1, \ldots, X^N$ *and* X^0 *are sections of* E, *then*

$$A^S = \frac{1}{2} \sum_{i,j=1}^{N} (a^{ij} \circ p) \, \mathbf{L}_{\bar{X}^i} \mathbf{L}_{\bar{X}^j} + \mathbf{L}_{\bar{X}^0} \tag{1.17}$$

    for $\bar{X}^j(y) = h_y(X^j(p(y)))$.

Proof. Lemma 1.4.3 ensures that (i) and (ii) determine $\bar{\delta}$ uniquely as a smooth operator on smooth 1-forms if it exists. On the other hand we can represent A as in (1.16) and define A^S by (1.17). It is straightforward to check that then δ^{A^S} satisfies (i) and (ii). $\qquad\qquad\square$

Example 1.4.5. . In Example 1.4.1 above we can take $A = \frac{d^2}{dx^2}$ on $M = \mathbf{R} \geqslant 0$, then the lift to $N = \mathbf{R}^2 - \{0\}$ along the given distribution in polar co-ordinates is $\frac{\partial^2}{\partial r^2}$. On the other hand if we changed p to be given by $p(u) = u^2$ we could use the same distribution S but the lift would be changed to $\frac{1}{2r} \frac{\partial^2}{\partial r^2}$.

Definition 1.4.6. When an operator B is along the vertical distribution ker$[Tp]$ we say B is **vertical**, and when there is a horizontal distribution such as $\{H_u : u \in N\}$ as given by Proposition 2.1.2 below and B is along that horizontal distribution we say B is **horizontal** .

Proposition 1.4.7. *Let* B *be a smooth diffusion operator on* N *and* $p : N \to M$ *any smooth map, then the following conditions are equivalent:*

(1) *The operator* B *is vertical.*

(2) *The operator* B *has an expression of the form of* $\sum_{j=1}^{m} a^{ij} \mathbf{L}_{Y^i} \mathbf{L}_{Y^j} + \mathbf{L}_{Y^0}$ *where* a^{ij} *are smooth functions and* Y^j *are smooth sections of the vertical tangent bundle of* TN.

(3) $B(f \circ p) = 0$ *for all* C^2 $f : M \to \mathbf{R}$.

Proof. (a). From (1) to (3) is trivial. From (3) to (1) note that every ϕ which vanishes on vertical vectors is a linear combination of elements of the form $fp^*(dg)$ for some smooth $g : M \to \mathbf{R}$ by Lemma 1.4.3. To show that $\mathcal{B}$ is vertical we only need to show that $\delta^{\mathcal{B}}(fp^*(dg)) = 0$. But $\mathcal{B}(g \circ p) = 0$ implies $\delta^{\mathcal{B}}(p^*(dg)) = 0$ and also $p^*(dg)\sigma^{\mathcal{B}}(p^*(dg)) = \frac{1}{2}\mathcal{B}(g \circ p)^2 - (g \circ p)\mathcal{B}(g \circ p) = 0$. By semi-ellipticity of $\mathcal{B}$, $\sigma^{\mathcal{B}}(p^*(dg)) = 0$. Thus assertion (1) follows since $\delta^{\mathcal{B}}(fp^*(dg)) = df\sigma^{\mathcal{B}}(p^*(dg)) + f \cdot \delta^{\mathcal{B}}(p^*(dg))$ from Proposition 1.2.1(1), and so (1) and (3) are equivalent.

Equivalence of (1) and (2) follows from Proposition 1.3.7. $\square$

Remark 1.4.8.   (1) If $\mathcal{B}$ is vertical, then by Proposition 1.2.1, for all C^2 functions f_1 on N and f_2 on M, $\mathcal{B}(f_1(f_2 \circ p)) = (f_2 \circ p)\mathcal{B}f_1$;

(2) If $\mathcal{B}$ and $\mathcal{B}'$ are both over a diffusion operator $\mathcal{A}$ of constant nonzero rank such that $\mathcal{A}$ is along the image of $\sigma^{\mathcal{A}}$, then $\mathcal{B} - \mathcal{B}'$ is not in general vertical, although $(\mathcal{B} - \mathcal{B}')(f \circ p) = 0$ for all C^2 function $f : M \to \mathbf{R}$, since it may not be semi-elliptic. For example take $p : \mathbf{R}^2 \to \mathbf{R}$ to be the projection $p(x, y) = x$ with $\mathcal{A} = \frac{\partial^2}{\partial x^2}$, $\mathcal{B} = \frac{\partial^2}{\partial x^2} + \frac{\partial^2}{\partial y^2}$. Let $\mathcal{B}' = \frac{\partial^2}{\partial x^2} + \frac{\partial^2}{\partial y^2} + \frac{\partial^2}{\partial x \partial y}$. Then $\mathcal{B}'$ is also over $\mathcal{A}$ but $\mathcal{B} - \mathcal{B}' = -\frac{\partial^2}{\partial x \partial y}$ is not vertical. In particular this shows that Proposition 1.4.7 would not hold without the assumption that $\mathcal{B}$ is a diffusion operator.

1.5   Notes

On symbols

What we have called the symbol of our diffusion operator is often called the **principal symbol**. In the semi-elliptic case it is related to the **energy density** used in Dirichlet form theory mainly for symmetric diffusions, and is also given in terms of the **carré du champ** or **squared gradient** operator Γ as used, for example, in Bakry-Emery theory, [3], by

$$\sigma^{\mathcal{L}}(df, dg) = \Gamma(f, g).$$

Chapter 2

Decomposition of Diffusion Operators

Consider again a smooth map $p : N \to M$ between smooth manifolds M and N and a lift $\mathcal{B}$ of a diffusion operator $\mathcal{A}$ on M.

Definition 2.0.1. In this situation we say that $\mathcal{B}$ **is over** $\mathcal{A}$, or that $\mathcal{A}$ and $\mathcal{B}$ are **intertwined** by p. In general a diffusion operator $\mathcal{B}$ on N is said to be **projectible** (over p), or p-*projectible*, if it is over some diffusion operator $\mathcal{A}$.

Recall that the pull-back $p^*\phi$ of a 1-form ϕ is defined by

$$p^*(\phi)_u = \phi_{p(u)}(Tp(-)) = (Tp)^*\phi_{p(u)}.$$

In local co-ordinates

$$p^*(\phi)_i = \sum_k \frac{\partial p^k}{\partial x_i}\phi_k \circ p.$$

For our map $p : N \to M$, a diffusion operator $\mathcal{B}$ is over $\mathcal{A}$ if and only if

$$\delta^{\mathcal{B}}\left(p^*\phi)\right) = (\delta^{\mathcal{A}}\phi)(p), \tag{2.1}$$

for all $\phi \in C^1 \wedge^1 T^*M$.

2.1 The Horizontal Lift Map

Lemma 2.1.1. *Suppose that $\mathcal{B}$ is over $\mathcal{A}$. Let $\sigma^{\mathcal{B}}$ and $\sigma^{\mathcal{A}}$ be respectively the symbols for $\mathcal{B}$ and $\mathcal{A}$. Then*

$$(T_up)\sigma^{\mathcal{B}}_u(T_up)^* = \sigma^{\mathcal{A}}_{p(u)}, \qquad \forall u \in N, \tag{2.2}$$

i.e. the following diagram is commutative:

K.D. Elworthy et al., *The Geometry of Filtering*, Frontiers in Mathematics, DOI 10.1007/978-3-0346-0176-4_2, © Springer Basel AG 2010

$$
\begin{array}{ccc}
T_u^*N & \xrightarrow{\;\sigma_u^{\mathcal{B}}\;} & T_uN \\[2pt]
\big\downarrow{\scriptstyle (T_up)^*} & & \big\downarrow{\scriptstyle T_up} \\[2pt]
T_{p(u)}^*M & \xrightarrow{\;\sigma_{p(u)}^{\mathcal{A}}\;} & T_{p(u)}M.
\end{array}
$$

Proof. Let f and g be two smooth functions on M. Then for $u \in N$, $x = p(u)$,

$$
\begin{aligned}
(df_x)\,\sigma_x^{\mathcal{A}}\,(dg_x) &= \frac{1}{2}\mathcal{A}(fg)(x) - \frac{1}{2}(f\mathcal{A}g)(x) - \frac{1}{2}(g\mathcal{A}f)(x) \\
&= \frac{1}{2}\mathcal{B}\,((fg)\circ p)\,(u) - \frac{1}{2}f\circ p\mathcal{B}(g\circ p)(u) - \frac{1}{2}g\circ p\mathcal{B}(f\circ p)(u) \\
&= d\,(g\circ p)_u\,\sigma_u^{\mathcal{B}}\,(d\,(f\circ p)_u) \\
&= (dg\circ T_up)\,\sigma_u^{\mathcal{B}}\,(df\circ T_up),
\end{aligned}
$$

which gives the desired equality. $\qquad\qquad\qquad\qquad\qquad\qquad\qquad\qquad\quad\square$

For x in M, set $E_x := \mathrm{Image}[\sigma_x^{\mathcal{A}}] \subset T_xM$. If $\sigma^{\mathcal{A}}$ has constant rank, *i.e.* $\dim[E_x]$ is independent of x, then $E := \cup_x E_x$ is a smooth subbundle of TM.

Proposition 2.1.2. *Assume $\sigma^{\mathcal{A}}$ has constant rank and $\mathcal{B}$ is over $\mathcal{A}$. Then there is a unique, smooth, horizontal lift map $h_u : E_{p(u)} \to T_uN$, $u \in N$, characterised by*

$$
h_u \circ \sigma_{p(u)}^{\mathcal{A}} = \sigma_u^{\mathcal{B}}(T_up)^*. \tag{2.3}
$$

In particular

$$
h_u(v) = \sigma_u^{\mathcal{B}}\left((T_up)^*\alpha\right) \tag{2.4}
$$

where $\alpha \in T_{p(u)}^*M$ satisfies $\sigma_{p(u)}^{\mathcal{A}}(\alpha) = v$.

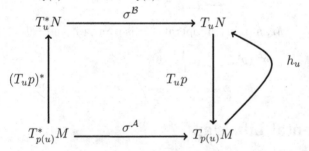

Proof. Clearly (2.4) implies (2.3) by Lemma 2.1.1 and so it suffices to prove h_u is well defined by (2.4). For this we only need to show $\sigma^{\mathcal{B}}((T_up)^*(\alpha)) = 0$ for every α in $\ker[\sigma_{p(u)}^{\mathcal{A}}]$. Now $\sigma^{\mathcal{A}}\alpha = 0$ implies that

$$
(Tp)^*(\alpha)\sigma^{\mathcal{B}}((Tp)^*\alpha) = 0,
$$

by Lemma 2.1.1. Considering σ^B as a semi-definite bilinear form this implies $\sigma^B_u(T_up)^*\alpha$ vanishes as required. □

Recall from Lemma 1.3.6 that the vertical distribution $\ker[Tp]$ is regular. Let $H_u = \text{Image}[h_u]$, the **horizontal subspace** at u, and $H = \sqcup_u H_u$. Set $F_u = (T_up)^{-1}[E_{p(u)}]$ so we have a splitting

$$F_u = H_u + VT_uN \tag{2.5}$$

where $VT_uN = \ker[T_uP]$ the 'vertical' tangent space at u to N.

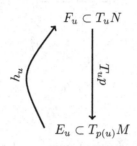

In the elliptic case p is a submersion, the vertical tangent spaces have constant rank, and $F := \sqcup_u F_u$ is a smooth subbundle of TN. In this case we have a splitting of TN, a **connection** in the terminology of Kolar-Michor-Slovak [57]. In general we will define a **semi-connection** on E to be a subbundle H_u of TN such that T_up maps each fibre H_u isomorphically to $E_{p(u)}$. In the equivariant case considered in Chapter 3 such objects are called E-connections by Gromov. For the case when $p : N \to M$ is the tangent bundle projection, or the orthonormal frame bundle, note that the "partial connections" as defined by Ge in [45] are rather different from the semi-connections we would have: they give parallel translations along E-horizontal paths which send vectors in E to vectors in E, and preserve the Riemannian metric of E, whereas the parallel transports of our semi-connections do not in general preserve the fibres of E, nor any Riemannian metric, and they act on all tangent vectors.

Lemma 2.1.3. *Assume σ^A has constant rank and $\mathcal{B}$ is over $\mathcal{A}$. For all $u \in N$ the image of σ^B_u is in F_u.*

Proof. Suppose $\alpha \in T^*_uN$ with $\sigma^B(\alpha) \notin F_u$. Then there exists k in the annihilator of $E_{p(u)}$ such that $k\left(T_up\,\sigma^B(\alpha)\right) \neq 0$. However

$$k\left(T_up\,\sigma^B(\alpha)\right) = \alpha\left(\sigma^B((T_up)^*(k))\right) = \alpha\,h_u\sigma^A_{p(u)}(k)$$

by Proposition 2.1.2; while $\sigma^A_{p(u)}(k) = 0$ because for all $\beta \in T^*_{p(u)}M$,

$$\beta\,\sigma^A_{p(u)}(k) = k\,\sigma^A_{p(u)}(\beta) = 0$$

giving a contradiction. □

Proposition 2.1.4. *Let $\mathcal{A}$ be a diffusion operator on M with $\sigma^{\mathcal{A}}$ of constant rank. For $i \in \{1, 2\}$, let $p^i : N^i \to M$ be smooth maps and $\mathcal{B}^i$ be diffusion operators on N^i over $\mathcal{A}$. Let $F : N^1 \to N^2$ be a smooth map with $p^2 \circ F = p^1$. Assume F intertwines $\mathcal{B}^1$ and $\mathcal{B}^2$. Let h^1, h^2 be the horizontal lift maps determined by $\mathcal{A}, \mathcal{B}^1$ and $\mathcal{A}, \mathcal{B}^2$. Then*

$$h^2_{F(u)} = T_u F(h^1_u), \qquad u \in N^1; \tag{2.6}$$

i.e. the diagram

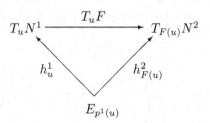

commutes for all $u \in N$.

Proof. Since F intertwines $\mathcal{B}^1$ and $\mathcal{B}^2$, Lemma 2.1.1 gives

$$\sigma^{\mathcal{B}^2}_{F(u)} = T_u F \circ \sigma^{\mathcal{B}^1}_u \circ (T_u F)^*.$$

Now take $\alpha \in T^*_{p^1(u)} M$ with $\sigma^{\mathcal{A}}_{p^1(u)}(\alpha) = v$, some given $v \in E_{p^1(u)}$. From (2.4)

$$\begin{aligned}
h^2_{F(u)}(v) &= \sigma^{\mathcal{B}^2}_{F(u)}((Tp^2)^*\alpha) \\
&= T_u F \circ \sigma^{\mathcal{B}^1}_u \circ (T_u F)^*(Tp^2)^*\alpha \\
&= T_u F \circ \sigma^{\mathcal{B}^1}_u (T_u p^1)^*\alpha \\
&= T_u h^1_u(v)
\end{aligned}$$

as required. $\square$

Definition 2.1.5. A diffusion operator $\mathcal{B}$ on N will be said to have **projectible symbol** for $p : N \to M$ if there exists a map $\eta : T^*M \to TM$ such that for all $u \in N$ the diagram

$$\begin{array}{ccc}
T^*_u N & \xrightarrow{\ \sigma^{\mathcal{B}}_u\ } & T_u N \\[4pt]
{\scriptstyle (T_u p)^*}\Big\uparrow & & \Big\downarrow{\scriptstyle T_u p} \\[4pt]
T^*_{p(u)} M & \xrightarrow[\ \eta_{p(u)}\ ]{} & T_{p(u)} M.
\end{array}$$

commutes, i.e. if $(T_u p)\sigma^{\mathcal{B}}_u (T_u p)^*$ depends only on $p(u)$.

In this case we also get a uniquely defined horizontal lift map as in Proposition 2.1.4 defined by equation (2.6) using η instead of the symbol of $\mathcal{A}$. This situation arises naturally in the standard non-linear filtering literature as described later, see Chapter 5.

Example 2.1.6. 1. Consider $N = \mathbf{R}^2$, $M = \mathbf{R}$, with $p(x,y) = x$. Take

$$\mathcal{B} = a(x)\frac{\partial^2}{\partial x^2} + 2b(x,y)\frac{\partial^2}{\partial x \partial y} + c(x,y)\frac{\partial^2}{\partial y^2}$$

so $\mathcal{A} = a(x)\frac{\partial^2}{\partial x^2}$. For semi-ellipticity of $\mathcal{B}$ we require $a(x)c(x,y) \geqslant b^2(x,y)$ and $a(x) \geqslant 0$ for all $(x,y) \in \mathbf{R}^2$. For the constant rank condition on $\mathcal{A}$ and non-triviality we will assume $a(x) > 0$ for all real x. Define $\gamma(x,y) = \frac{b(x,y)}{a(x)}$. Then the horizontal lift map, or more precisely its principal part, $h_u : \mathbf{R} \to \mathbf{R}^2$ at $u = (x,y)$, is given by

$$h_{(x,y)}(r) = (r, \gamma(x,y)r). \tag{2.7}$$

To check this satisfies the defining criterion equation (2.3) observe that with our definition, for $r \in \mathbf{R}$,

$$h_u \circ \sigma_{p(u)}^{\mathcal{A}} r = (\sigma_{p(u)}^{\mathcal{A}} r, \gamma(u)\sigma_{p(u)}^{\mathcal{A}} r) \tag{2.8}$$

$$= (a(x)r, b(x,y)r); \tag{2.9}$$

while

$$\sigma_u^{\mathcal{B}}(T_u p)^* r = \begin{pmatrix} a(x) & b(x,y) \\ b(x,y) & c(x,y) \end{pmatrix} \begin{pmatrix} r \\ 0 \end{pmatrix} \tag{2.10}$$

$$= (a(x)r, b(x,y)r). \tag{2.11}$$

2. More generally take $N = \mathbf{R}^n \times \mathbf{R}$, $M = \mathbf{R}^n$, with p the projection. Now take

$$(\mathcal{B}f)(x,y) = \sum_{i,j=1}^{n} a^{i,j}(x)\frac{\partial^2}{\partial x_i \partial x_j}f + \sum_{k=1}^{n} b^k(x,y)\frac{\partial^2}{\partial x^k \partial y}f + c(x,y)\frac{\partial^2}{\partial y^2}f$$

where $a = (a_{i,j})$ is a symmetric $n \times n$-matrix-valued function, $b = (b^1, \ldots, b^n)$ takes values in $\mathbf{R}^n$ and c is real-valued.

Then $\mathcal{B}$ lies over $\mathcal{A}$ for

$$\mathcal{A} = \sum_{i,j=1}^{n} a^{i,j}(x)\frac{\partial^2}{\partial x^i \partial x^j}.$$

For $\mathcal{A}$ to be semi-elliptic with symbol of constant rank we require $a(x)$ to be positive semi-definite for all $x \in M$ and to have constant rank. It is easy to see that, assuming this, $\mathcal{B}$ will be semi-elliptic if and only if for all $\xi \in \mathbf{R}^n$ we have

$$\langle b(x,y)\xi, \xi\rangle^2 \leqslant c(x,y)\langle a(x)\xi, \xi\rangle \tag{2.12}$$

or equivalently, as matrices,

$$b(x,y)^t b(x,y) \leqslant c(x,y)a(x) \tag{2.13}$$

for all $(x,y) \in N$.

For this to hold we see that $b(x,y)$ must always lie in the image of $a(x)$, otherwise there will be some ξ orthogonal to the image of $a(x)$ for which $\langle b(x,y)\xi, \xi\rangle^2 > 0$ contradicting condition (2.12).

Moreover since $a(x)$ is symmetric its kernel is orthogonal to its image, and so if v is in its image, $\langle a(x)^{-1}b, v\rangle$ is well defined. In fact it is simply the inner product $\langle b, v\rangle_x$ of v and b in the metric on E determined by $\mathcal{A}$. In this notation the horizontal lift map $h_{(x,y)} : E_x \to N \times (\mathbf{R}^n \times \mathbf{R})$ is given by

$$h_{(x,y)}(v) = ((x,y),(v,\langle b,v\rangle_x)). \tag{2.14}$$

3. More generally if $\mathcal{A}(x)$ is given by a positive definite matrix $(m+p) \times (m+p)$ matrix A of rank m, $B(x,y)$ is an $(m+p) \times q$ matrix with $B(x,y)$ in the image of $A(x)$, and $C(x,y)$ a $q \times q$ matrix. We have a horizontal lifting map $u \in Image(A(x)) \to (u, B^T(x,y)A^{-1}(x)u)$.

Example 2.1.7 (Coupling of diffusion operators). Consider diffusion operators $\mathcal{A}^1$ and $\mathcal{A}^2$ on manifolds M^1 and M^2. Take $N = M_1 \times M_2$ with p_1 and p_2 the corresponding projections. A diffusion operator $\mathcal{B}$ on N is a *coupling* of $\mathcal{A}^1$ with $\mathcal{A}^2$ if $\mathcal{B}$ and A^j are intertwined by p_j for $j = 1, 2$. If so it is easy to see that there is a bilinear $\Gamma^{\mathcal{B}} : T^*M^1 \times T^*M^2 \to \mathbf{R}$ such that

$$\mathcal{B}(f \otimes g)(x,y) = \mathcal{A}^1(f)(x)g(y) + f(x)\mathcal{A}^2(g)(y) + \Gamma^{\mathcal{B}}((df)_x, (dg)_y) \tag{2.15}$$

where $f \otimes g : M^1 \times M^2 \to \mathbf{R}$ denotes the map $(x,y) \mapsto f(x)g(y)$ and f, g are C^2. Note that the symbol $\sigma^{\mathcal{B}} : T^*M_1 \times T^*M_2 \to TM_1 \times TM_2$ is given by

$$\sigma^{\mathcal{B}}_{(x,y)}(\ell_1, \ell_2) = \left(\sigma^{\mathcal{A}^1}_x(\ell_1) + \sigma^{1,2}_{(x,y)}(\ell_2), \sigma^{\mathcal{A}^2}_x(\ell_2) + \sigma^{2,1}_{(x,y)}(\ell_1)\right),$$
$$\ell_1 \in T^*_x M_1, \ell_2 \in T^*_y M_2 \tag{2.16}$$

where $\sigma^{1,2}_{(x,y)} : T^*_y M_2 \to T_x M_1$ and $\sigma^{2,1}_{(x,y)} : T^*_x M_1 \to T_y M_2$ are defined by

$$\ell_1 \sigma^{1,2}_{(x,y)}(\ell_2) = \frac{1}{2}\Gamma^{\mathcal{B}}(\ell_1, \ell_2) = \ell_2 \sigma^{2,1}_{(x,y)}(\ell_1).$$

Now take $\mathcal{A} = \mathcal{A}^1$ and $p = p_1$ and assume $\sigma^{\mathcal{A}^1}$ has constant rank.

Lemma 2.1.8. *For $(x, y) \in M_1 \times M_2$:*

1. $\sigma^{1,2}_{(x,y)} = (\sigma^{2,1}_{(x,y)})^* : T^*_y M_2 \to T_x M_1,$

2. $\sigma^{1,2}_{(x,y)}$ *has image in E_x.*

Proof. The first assertion is immediate from the definitions. For the second we use the semi-ellipticity of $\mathcal{B}$ with the argument in the proof of the existence of a horizontal lift, Proposition 2.1.2, to see that if $\ell \in \ker \sigma^A_x$, then, for all $y \in M_2$ we have $\sigma^{\mathcal{B}}(T_{(x,y)}p)^*(\ell) = 0$. This means that $\ell \in \ker \sigma^{2,1}_{(x,y)}$. By the first part this implies that $\ell \sigma^{1,2}_{(x,y)}(\tilde{\ell}) = 0$ for all $\tilde{\ell} \in T^*_y M_2$. Thus $\ell \in \ker \sigma^A_x$ implies that ℓ annihilates the image of $\sigma^{1,2}_{(x,y)}$. On the other hand since $\mathcal{A}$ is symmetric $\ell \in \ker \sigma^A_x$ if and only if ℓ annihilates E_x. $\qquad\square$

By the lemma we can use the Riemannian metric on E induced by the symbol of $\mathcal{A}$ to define the adjoint $(\sigma^{1,2}_{x,y})^\# : E_x \to T_y M_2$ of $\sigma^{1,2}_{(x,y)}$. We claim that the horizontal lift of the semi-connection induced by our coupling is given by

$$h_{(x,y)}(v) = \left(v, (\sigma^{1,2}_{x,y})^\#(v)\right) \in T_x M_1 \times T_y M_2 \quad v \in E_x. \qquad (2.17)$$

To check this, first note that from Proposition 2.1.2 we know

$$h_{(x,y)}(v) = \left(v, \sigma^{2,1}_{(x,y)}((\sigma^A_x)^{-1}(v))\right).$$

Next take $\ell_1 \in T^*_x M_1$ and $\ell_2 \in T^*_y M_2$. Write σ_x for σ^A considered as a map from $T^*_x M_1 \to E_x$. Then our claim follows from:

$$\ell_2[(\sigma^{1,2}_{(x,y)})^\# \sigma_x(\ell_1)] = \ell_1 \sigma^*_x(\ell_2 \circ (\sigma^{1,2}_{(x,y)})^\#)$$
$$= \langle \ell_1|_{E_x}, \ell_2 \circ (\sigma^{1,2}_{(x,y)})^\# \rangle_{E^*_x}$$
$$= \ell_2 \sigma^{2,1}_{(x,y)}(\ell_1).$$

2.2 Lifts of Cohesive Operators and The Decomposition Theorem

A diffusion generator $\mathcal{L}$ on a manifold is said to be **cohesive** if

(i) $\sigma^{\mathcal{L}}_x$, $x \in X$, has constant non-zero rank and

(ii) $\mathcal{L}$ is along the image of $\sigma^{\mathcal{L}}$.

Remark 2.2.1. From Theorem 2.1.1 in Elworthy-LeJan-Li [36] we see that if the rank of $\sigma^{\mathcal{L}}_x$ is bigger than 1 for all x, then $\mathcal{L}$ is cohesive if and only if it has a representation

$$\mathcal{L} = \frac{1}{2} \sum_{j=1}^{m} \mathbf{L}_{X^j} \mathbf{L}_{X^j}$$

where $E_x = \text{span}\{X^1(x), \ldots X^m(x)\}$ has constant rank.

Let H be the horizontal distribution of the semi-connection determined by a cohesive diffusion generator $\mathcal{A}$. We can now define the **horizontal lift** of $\mathcal{A}$ to be the diffusion generator $\mathcal{A}^H$ on N given by Proposition 1.4.4.

The equivalence of (i) and (ii) in the following proposition shows that $\mathcal{A}^H$ can be characterised independently of any semi-connection.

Proposition 2.2.2. *Let $\mathcal{B}$ be a smooth diffusion operator on N over $\mathcal{A}$ with $\mathcal{A}$ cohesive. The following are equivalent:*

(i) $\mathcal{B} = \mathcal{A}^H$.

(ii) *$\mathcal{B}$ is cohesive and $T_u p$ is injective on the image of $\sigma_u^{\mathcal{B}}$ for all $u \in N$.*

(iii) *$\mathcal{B}$ can be written as*

$$\mathcal{B} = \frac{1}{2} \sum_{j=1}^{m} \mathbf{L}_{\tilde{X}^j} \mathbf{L}_{\tilde{X}^j} + \mathbf{L}_{\tilde{X}^0}$$

where $\tilde{X}^0, \ldots, \tilde{X}^m$ are smooth vector fields on N lying over smooth vector fields $X^0, \ldots, X^m$ on M, i.e. $T_u p(\tilde{X}^j(u)) = X^j(p(u))$ for $u \in N$ for all j.

Proof. If (i) holds, take smooth $X^1, \ldots, X^m$ with $\mathcal{A} = \frac{1}{2} \sum_{j=1}^{m} \mathbf{L}_{X^j} \mathbf{L}_{X^j} + \mathbf{L}_{X^0}$, by Proposition 1.3.7, and set $\tilde{X}^j(u) = h_u X^j(p(u))$ to see (iii) holds. Clearly (iii) implies (ii) and (ii) implies (i), so the three statements are equivalent. □

Definition 2.2.3. If condition (ii) of the proposition holds we say that $\mathcal{B}$ *has no vertical part.*

Recall that if S is a distribution in TN, then S^0 denotes the set of annihilators of S in $T^* N$.

Lemma 2.2.4. *For $\ell \in H_u^0$ and $k \in (V_u TN)^0$, some $u \in N$ we have:*

A. $\ell \sigma^{\mathcal{B}}(k) = 0$,

B. $\sigma^{\mathcal{B}}(k) = \sigma^{\mathcal{A}^H}(k)$,

C. $\sigma^{\mathcal{A}^H}(\ell) = 0$.

In particular H_u is the orthogonal complement of $VT_u N \cap \text{Image}(\sigma_u^{\mathcal{B}})$ in $\text{Image}(\sigma_u^{\mathcal{B}})$ with its inner product induced by $\sigma_u^{\mathcal{B}}$.

Proof. Set $x = p(u)$. For part A and part B it suffices to take $k = \phi \circ T_u p$ some $\phi \in T_x^* M$. Then by (2.3), $\sigma_u^{\mathcal{B}}(\phi \circ T_u p) = h_u \circ \sigma_x^{\mathcal{A}}(\phi)$ giving part A, and also part B by Proposition 1.4.4 (iii)(a) since $\phi = h_u^*(\phi \circ T_u p)$. Part C comes directly from Proposition 1.4.4 (iii)(a). □

Theorem 2.2.5. *For $\mathcal{B}$ over $\mathcal{A}$ with $\mathcal{A}$ cohesive there is a unique decomposition*

$$\mathcal{B} = \mathcal{B}^1 + \mathcal{B}^V$$

where $\mathcal{B}^1$ and $\mathcal{B}^V$ are smooth diffusion generators with $\mathcal{B}^V$ vertical and $\mathcal{B}^1$ over $\mathcal{A}$ having no vertical part. In this decomposition $\mathcal{B}^1 = \mathcal{A}^H$, the horizontal lift of $\mathcal{A}$ to H.

Proof. Set $\mathcal{B}^V = \mathcal{B} - \mathcal{A}^H$. To see that $\mathcal{B}^V$ is semi-elliptic take $u \in N$ and observe that any element of $T_u^* N$ can be written as $\ell + k$ where $\ell \in H_u^0$ and $k \in (VT_uN)^0$ by Lemma 2.2.4 and

$$(\ell + k)\sigma^{\mathcal{B}}(\ell + k) = \ell\sigma^{\mathcal{B}}(\ell) \geqslant 0.$$

Since $\mathcal{B}^V(f \circ p) = 0$ any $f \in C^2(M; \mathbf{R})$, Proposition 1.4.7 implies $\mathcal{B}^V$ is vertical. Uniqueness holds since the semi-connections determined by $\mathcal{B}$ and $\mathcal{B}'$ are the same by Remark 1.3.4(i) applied to $\mathcal{B}^V$ and so by Proposition 2.2.2 we must have $\mathcal{B}^1 = \mathcal{A}^H$. $\square$

Remark 2.2.6. For p a Riemannian submersion and $\mathcal{B}$ the Laplacian, Berard-Bergery and Bourguignon [9] define $\mathcal{B}^V$ directly by $\mathcal{B}^V f(u) = \Delta_{N_x}(f|_{N_x})(u)$ for $x = p(u)$ and $N_x = p^{-1}(x)$ with Δ_{N_x} the Laplace-Beltrami operator of N_x.

Definition 2.2.7. For a smooth $p : N \to M$ vector fields $\tilde{A}$ and A on N and M respectively are said to be p-**related** if

$$T_u p\left(\tilde{A}(u)\right) = A(p(u)) \quad \text{for all} \quad u \in N.$$

Remark 2.2.8. When $\mathcal{A}$ and $\mathcal{B}$ are given in Hörmander forms using p-related vector fields, another decomposition of $\mathcal{B}$ is described in the Appendix, Section 9.5 Remark 9.5.

Example 2.2.9. Returning to Example 2.1.6 for $p : \mathbf{R}^2 \to \mathbf{R}$ the projection and

$$\mathcal{B} = a(x)\frac{\partial^2}{\partial x^2} + 2b(x,y)\frac{\partial^2}{\partial x \partial y} + c(x,y)\frac{\partial^2}{\partial y^2}$$

the decomposition is given by

$$\mathcal{B} = a(x)\left(\frac{\partial}{\partial x} + h(x,y)\frac{\partial}{\partial y}\right)^2 + d(x,y)\frac{\partial^2}{\partial y^2}$$

where as in Example 2.1.6 $h(x,y) = \frac{b(x,y)}{a(x)}$ while $d(x,y) = c(x,y) - \frac{b(x,y)^2}{a(x)}$. This follows from the fact that from Example 2.1.6 we know that the horizontal subspace at (x,y) is just $\{(r, h(x,y)r) : r \in \mathbf{R}\}$. This means that the first term in our decomposition is horizontal (while clearly a lift of $\mathcal{A}$); the second term is clearly vertical.

Example 2.2.10. Take $N = S^1 \times S^1$ and $M = S^1$ with p the projection on the first factor. Let

$$\mathcal{B} = \frac{1}{2}\Big(\frac{\partial^2}{\partial x^2} + \frac{\partial^2}{\partial y^2}\Big) + \tan\alpha \frac{\partial^2}{\partial x \partial y}.$$

Here $0 < \alpha < \frac{\pi}{4}$ so that $\mathcal{B}$ is elliptic. Then $\mathcal{A} = \frac{1}{2}\frac{\partial^2}{\partial x^2}$,

$$\mathcal{B}^V = \frac{1}{2}(1 - (\tan\alpha)^2)\frac{\partial^2}{\partial y^2}, \quad \mathcal{A}^H = \frac{1}{2}\Big(\frac{\partial^2}{\partial x^2} + (\tan\alpha)^2\frac{\partial^2}{\partial y^2}\Big) + \tan\alpha \frac{\partial^2}{\partial x \partial y}.$$

This is easily checked since, with this definition $\mathcal{A}^H$ has Hörmander form

$$\mathcal{A}^H = \frac{1}{2}\Big(\frac{\partial}{\partial x} + \tan\alpha \frac{\partial}{\partial y}\Big)^2$$

and so is a diffusion operator which has no vertical part. Also $\mathcal{B}^V$ is clearly vertical and elliptic. Note that this is another example of a Riemannian submersion: several more of a similar type can be found in [9]. In this case the horizontal distribution is integrable and if α is irrational the foliation it determines has dense leaves.

Example 2.2.11. Take $N = H$, the first Heisenberg group with *Heisenberg group* action as in Example 1.3.3,

$$(x, y, z) \cdot (x', y', z') = \Big(x + x', y + y', z + z' + \frac{1}{2}(xy' - yx')\Big).$$

As before let X, Y, Z be the left-invariant vector fields which give the standard basis for $\mathbf{R}^3$ at the origin so that as operators:

$$X(x, y, z) = \frac{\partial}{\partial x} - \frac{1}{2}y\frac{\partial}{\partial z}, \qquad Y(x, y, z) = \frac{\partial}{\partial y} + \frac{1}{2}x\frac{\partial}{\partial z},$$

$$Z(x, y, z) = \frac{\partial}{\partial z}.$$

Take $\mathcal{B}$ to be half the sum of the squares of X, Y, and Z. This is half the left invariant Laplacian:

$$\mathcal{B} = \frac{1}{2}\left(\frac{\partial^2}{\partial x^2} + \frac{\partial^2}{\partial y^2} + (1 + \frac{1}{4}(x^2 + y^2))\frac{\partial^2}{\partial z^2} + (x\frac{\partial^2}{\partial y \partial z} - y\frac{\partial^2}{\partial x \partial z})\right).$$

Take $M = \mathbf{R}^2$ and $p : \mathbf{R}^3 \to \mathbf{R}^2$ to be the projection on the first 2 coordinates. Then the horizontal lift map, induced by $(\mathcal{A}, \mathcal{B})$, from $\mathbf{R}^2$ to H is

$$h_{(x,y,z)} : (u, v) \mapsto (u, v, \frac{1}{2}(xv - yu))$$

and

$$\mathcal{A} = \frac{1}{2}\Big(\frac{\partial^2}{\partial x^2} + \frac{\partial^2}{\partial y^2}\Big), \qquad \mathcal{A}^H = \frac{1}{2}(X^2 + Y^2),$$

$$\mathcal{B}^V = \frac{1}{2}Z^2 = \frac{1}{2}\frac{\partial^2}{\partial z^2}.$$

The decomposition of the operator is just the completion of squares. This leads back to the canonical left-invariant horizontal vector fields:

$$2\mathcal{B} = \frac{\partial^2}{\partial x^2} + \frac{\partial^2}{\partial y^2} + (1 + \frac{1}{4}(x^2 + y^2))\frac{\partial^2}{\partial z^2}$$
$$+ (x\frac{\partial^2}{\partial z \partial y} - y\frac{\partial^2}{\partial z \partial x})$$
$$= (\frac{\partial}{\partial x} - \frac{y}{2}\frac{\partial}{\partial z})^2 + (\frac{\partial}{\partial y} + \frac{x}{2}\frac{\partial}{\partial z})^2 + \frac{\partial^2}{\partial z^2}.$$

This philosophy we maintain throughout the book.

Note that the horizontal lift $\tilde{\sigma}$, of a smooth curve $\sigma : [0, T] \to M$ with $\sigma(0) = 0$, is given by

$$\tilde{\sigma}(t) = \left(\sigma^1(t), \sigma^2(t), \frac{1}{2}\int_0^t (\sigma^1(t)d\sigma^2(t) - \sigma^2(t)d\sigma^1(t))\right). \tag{2.18}$$

Thus the "vertical" component of the horizontal lift is the area integral of the curve. Equation (2.18) remains valid for the horizontal lift of Brownian motion on $\mathbf{R}^2$, or more generally for any continuous semi-martingale, provided it is interpreted as a Stratonovich equation (or equivalently an Itô equation in the Brownian motion case). This example is also that of a Riemannian submersion. In this case the horizontal distributions are not integrable. Indeed the Lie brackets satisfy $[X, Y] = Z$ and *Hörmander's condition* for hypoellipticity: a diffusion operator $\mathcal{L}$ satisfies Hörmander's condition if for some (and hence all) Hörmander form representation such as in equation (1.14) the vector fields $Y^1, \ldots, Y^m$ together with their iterated Lie brackets span the tangent space at each point of the manifold. For an enjoyable discussion of the Heisenberg group and the relevance of this example to "Dido's problem" see [79]. See also [5],[11], and [49].

Example 2.2.12. For nontrivial connections on the Heisenberg group discussed above, consider $\mathcal{A} = \frac{1}{2}(\frac{\partial^2}{\partial x^2} + \frac{\partial^2}{\partial y^2})$ as before, and for real-valued functions r_1, r_2, γ with $\gamma > r_1^2 + r_2^2$,

$$\mathcal{B}_{r_1,r_2} = \frac{1}{2}(\frac{\partial^2}{\partial x^2} + \frac{\partial^2}{\partial y^2}) + r_1\frac{\partial^2}{\partial x \partial z} + r_2\frac{\partial^2}{\partial y \partial z} + \frac{1}{2}\gamma\frac{\partial^2}{\partial^2 z}.$$

The horizontal lift map is:

$$h_{(x,y,z)}(u, v) = (u, v, r_1 u + r_2 v),$$

$$X_1 := h_u(\frac{\partial}{\partial x}) = (\frac{\partial}{\partial x} + r_1\frac{\partial}{\partial z}), \qquad X_2 := h_u(\frac{\partial}{\partial y}) = (\frac{\partial}{\partial y} + r_2\frac{\partial}{\partial z})$$

and

$$\mathcal{B}_{r_1,r_2} = \frac{1}{2}(X_1^2 + X_2^2) + \frac{1}{2}(\gamma - r_1^2 - r_2^2)\frac{\partial^2}{\partial^2 z}.$$

Example 2.2.13. Consider $N = \mathbf{R}^3 \times \mathbf{R}^3$ with coordinates

$$u = (v_1, v_2, v_3; w_1, w_2, w_3)(u).$$

Take $M = \mathbf{R}^3$ and $p(u) = (v_1, v_2, w_3)$. On N, for a fixed $\alpha \in \mathbf{R}$, consider the operator

$$\mathcal{B} = \sum_{1}^{3} \left(\frac{1}{2} \frac{\partial^2}{\partial v_i^2} + \frac{1}{2} \alpha^2 (v_{i+1}^2 + v_{i+2}^2) \frac{\partial^2}{\partial w_i^2} + \alpha (v_i \frac{\partial}{\partial v_{i+1}} - v_{i+1} \frac{\partial}{\partial v_i}) \frac{\partial}{\partial w_{i+2}} \right),$$

where the suffixes are to be taken modulo 3.

It projects by p on the operator

$$\mathcal{A} = \frac{1}{2} (\frac{\partial^2}{\partial v_1^2} + \frac{\partial^2}{\partial v_2^2}) + \frac{1}{2} \alpha^2 (v_1^2 + v_2^2) \frac{\partial^2}{\partial w_3^2} + \alpha (v_1 \frac{\partial}{\partial v_2} - v_2 \frac{\partial}{\partial v_1}) \frac{\partial}{\partial w_3}.$$

Note that $\mathcal{B} = \frac{1}{2} \sum_{1}^{3} X_i^2$ and $\mathcal{A} = \frac{1}{2} (Y_1^2 + Y_2^2)$ with

$$X_i = \frac{\partial}{\partial v_i} + \alpha (v_{i+2} \frac{\partial}{\partial w_{i+1}} - v_{i+1} \frac{\partial}{\partial w_{i+2}})$$

and

$$Y_1 = \frac{\partial}{\partial v_1} - \alpha v_2 \frac{\partial}{\partial w_3}, \tag{2.19}$$

$$Y_2 = \frac{\partial}{\partial v_2} + \alpha v_1 \frac{\partial}{\partial w_3}. \tag{2.20}$$

The horizontal lift is determined by the identities: $\mathcal{A}^H = \frac{1}{2}(X_1^2 + X_2^2)$, $\mathcal{B}^V = \frac{1}{2} X_3^2$ and $Y_i^H = X_i$, $i = 1, 2$.

Note that N is a group with multiplication given by $u \cdot u' = u''$, with $v_i(u'') = v_i(u) + v_i(u')$ and

$$w_{i+2}(u'') = w_{i+2}(u) + \alpha (v_i(u) v_{i+1}(u') - v_{i+1}(u) v_i(u')),$$

and p is a homomorphism of N onto the Heisenberg group. The diffusion generators $\mathcal{B}$ and $\mathcal{A}$ are invariant under left multiplication.

Recall that $F \equiv \sqcup_u F_u = \cup_u (T_u p)^{-1}[E_{p(u)}]$, we can now strengthen Lemma 2.1.3 which states that Image$[\sigma_u^{\mathcal{B}}] \subset F_u$.

Corollary 2.2.14. *If $\mathcal{B}$ is over $\mathcal{A}$ with $\mathcal{A}$ cohesive, then $\mathcal{B}$ is along F.*

Proof. Since $H_u \in F_u$ and $VT_u N \subset F_u$ both $\mathcal{B}^1$ and $\mathcal{B}^V$ are along F. □

2.3 The Lift Map for SDEs and Decomposition of Noise

Let us consider the horizontal lift connection in more detail when $\mathcal{B}$ and $\mathcal{A}$ are given by stochastic differential equations. For this write $\mathcal{A}$ and $\mathcal{B}$ in Hörmander form corresponding to factorisations $\sigma_x^{\mathcal{A}} = X(x)X(x)^*$ and $\sigma_x^{\mathcal{B}} = \tilde{X}(x)\tilde{X}(x)^*$ for

$$X(x) : \mathbf{R}^m \to T_x M, \qquad x \in M,$$

$$\tilde{X}(u) : \mathbf{R}^{\tilde{m}} \to T_u N, \qquad u \in N.$$

Then $X(x)$ maps onto E_x for each $x \in M$. Define $Y_x : E_x \to \mathbf{R}^m$ to be its right inverse: $Y_x = Y(x) = \left[X(x)\big|_{\ker X(x)^\perp} \right]^{-1}$.

Lemma 2.3.1. *For each $u \in N$ there is a unique linear $\ell_u : \mathbf{R}^m \to \mathbf{R}^{\tilde{m}}$ such that $\ker \ell_u = \ker X(p(u))$ and the diagram*

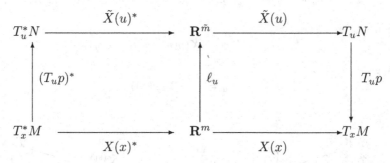

commutes, for $x = p(u)$, i.e. $\sigma_x^{\mathcal{A}} = T_u p \circ \sigma_x^{\mathcal{B}} (T_u p)^*$ and $X(x) = T_u p \circ \tilde{X}(u) \circ \ell_u$. In particular the horizontal lift map is given by $h_u = \tilde{X}(u)\ell_u Y(p(u))$.

Proof. The larger square commutes by Lemma 2.1.1. For the rest we need to construct ℓ_u. It suffices to define ℓ_u on $[\ker X(x)]^\perp$. Note that $[\ker X(x)]^\perp =$ Image $X(x)^*$ in $\mathbf{R}^m$. We only have to show that $\alpha \in \ker X(x)^*$ implies

$$\tilde{X}(u)^* (T_u p)^* \alpha = 0.$$

In fact for such α the proof in Proposition 2.1.2 is valid and therefore $(T_u p)^* \alpha \in \ker \sigma_u^{\mathcal{B}}$. However since $\tilde{X}(u)$ is injective on the image of $\tilde{X}(u)^*$ we see $\ker \sigma_u^{\mathcal{B}} = \ker \tilde{X}(u)^*$. Thus ℓ_u is defined with $\ker \ell_u = \ker X(x)$ and such that the left-hand square of the diagram commutes. Since the perimeter commutes it is easy to see from the construction of ℓ_u that the right-hand side also commutes. The uniqueness of ℓ_u with kernel equal to that of $X(x)$ is therefore clear since, on $[\ker X(x)]^\perp$, we require $\ell_u(e)$ to be $\tilde{X}(u)^* (T_u p)^*$ for any $\alpha \in T_x^* M$ with $X(x)^* \alpha = e$. $\square$

From now on assume that $X(x)$ has rank independent of $x \subset M$. This ensures that ℓ_u is smooth in $u \in N$. Also assume that $A(x) \in E_x$ for all $x \in M$, i.e. that $\mathcal{A}$ is cohesive. This is needed when we wish to consider the horizontal lift $\mathcal{A}^H$ of $\mathcal{A}$.

The horizontal lift of $X(x)$, which can be used to construct a Hörmander form representation of $\mathcal{A}^H$, as in Theorem 2.2.5 and Theorem 3.2.1 below is given by:

$$X^H(u) : \mathbf{R}^m \to T_u P,$$

$$X^H(u) = h_u X(u) = \tilde{X}(u)\ell_u$$

since $Y_x X(x)$ is the projection onto $\ker X(x)^\perp$.

Now for $x \in M$ let $K(x)$ be the orthogonal projection of $\mathbf{R}^m$ onto the kernel of $X(x)$ and $K^\perp(x)$ the projection onto $[\ker X(x)]^\perp$, so

$$K^\perp(x) = Y(x)X(x). \tag{2.21}$$

Consider the special case that $\tilde{m} = m$ and also that $\tilde{X}$ and X are p-**related**, i.e.

$$T_u p(\tilde{X}(u)e) = X(p(u))e, \qquad u \in N, e \in \mathbf{R}^m.$$

Then

$$\ell_u = Y(p(u))X(p(u)) = K^\perp(p(u))$$

giving

$$h_u = \tilde{X}(u)Y(p(u)) : \tag{2.22}$$

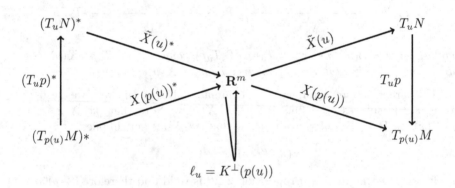

2.3.1 Decomposition of Stratonovich SDE's

Suppose we have an SDE on N:

$$du_t = \tilde{X}(u_t) \circ dB_t + \tilde{A}(u_t)\, dt \tag{2.23}$$

where $\tilde{X}$ is as above and p-related to X on M while the vector field $\tilde{A}$ is p-related to a vector field A on M. Thus if $x_t = p(u_t)$ we have

$$dx_t = X(x_t) \circ dB_t + A(x_t)\, dt. \tag{2.24}$$

Then we can decompose our SDE for u_t by:

$$du_t = \tilde{X}(u_t)K^\perp(p(u_t)) \circ dB_t + \tilde{A}(u_t) \, dt + \tilde{X}(u_t)K(p(u_t)) \circ dB_t$$

$$= X^H(u_t) \circ \, dB_t + A^H(u_t) \, dt + \tilde{X}(u_t)K(p(u_t)) \circ dB_t + \left(\tilde{A}(u_t) - A^H(u_t)\right) dt$$

$$= h_{u_t} \circ dx_t + \tilde{X}(u_t)K(p(u_t)) \circ dB_t + \left(\tilde{A}(u_t) - A^H(u_t)\right) dt.$$

We shall come back to such decompositions in Sections 4.8, 5.4, and 8.2.

2.3.2 Decomposition of the noise and Itô SDE's

The decomposition of our SDE on N described above is closely related to the decomposition of the noise of an SDE into essential and redundant components. This was first described in [38] with a more general discussion in [36]. The latter allowed for infinite dimensional noise and incomplete SDE. Here we will review the situation for our SDE (2.24) assuming it is complete, and X has constant rank with $A(x) \in E_x$ for all $x \in M$.

The projections K and $K^\perp$ determine metric connections on the subbundles $\ker X$ and $\ker^\perp X$ of the trivial bundle $\mathbf{R}^m$. Writing $\mathbf{R}^m- = \ker X \oplus \ker^\perp X$ we have the direct sum connection on $\mathbf{R}^m$. Let $/\!/_t \in O(m)$ for $t \geqslant 0$, be the corresponding parallel translation along the sample paths of the solution to our SDE (2.24) starting at a given point x_0 of M. Then $/\!/_t[\ker X(x_0)] = \ker X(x_t)]$ and $/\!/_t[\ker^\perp X(x_0)] = \ker^\perp X(x_t)$. Define processes $\{B_t^e\}_{t \geqslant 0}$ and $\{\beta_t\}_{t \geqslant 0}$ by

$$B_t^e = \int_0^t \tilde{/\!/}_s^{-1} K(x_s) \, dB_s \tag{2.25}$$

$$\beta_t = \int_0^t \tilde{/\!/}_s^{-1} K^\perp(x_s) \, dB_s. \tag{2.26}$$

Then:

1. $\{B_t^e\}_{t \geqslant 0}$ and $\{\beta_t\}_{t \geqslant 0}$ are independent Brownian motions, on $\ker^\perp X(x_0)$ and $\ker X(x_0)$ respectively;

2. The filtration of $\{B_t^e\}_{t \geqslant 0}$ is the same as that of $\{x_t\}_{t \geqslant 0}$;

3. $dB_t = \tilde{/\!/}_t \, dB_t^e + \tilde{/\!/}_t \, d\beta_t$.

The process $\beta_\cdot$ is the *redundant noise* and B^e, sometimes denoted by $\tilde{B}_\cdot$, the *relevant* or *essential* noise.

Suppose now that $M = \mathbf{R}^n$ and $N = \mathbf{R}^k$, and that we have an Itô SDE

$$du_t = \tilde{X}(u_t) \, dB_t + \tilde{A}(u_t) \, dt$$

on $\mathbf{R}^k$ whose solutions are such that if $x_t := p(u_t)$ for $t \geqslant 0$, then

$$dx_t = X(x_t) \, dB_t + A(x_t) \, dt \tag{2.27}$$

where X is p-related to $\tilde{X}$. Then from above we have:

$$du_t = \tilde{X}(u_t)/\!/_t \, dB_t^e + \tilde{A}(u_t) \, dt + \tilde{X}(u_t)/\!/_t \, d\beta_t$$
$$= X^H(u_t) \, dB_t + A^H(u_t) \, dt$$
$$+ \left(\tilde{A}(u_t) - A^H(u_t) \right) \, dt + \tilde{X}(u_t)/\!/_t \, d\beta_t.$$

However note that the solutions to $dz_t = X^H(z_t) \, dB_t + A^H(z_t) \, dt$ will not in general be horizontal lifts of solutions to equation (2.27) and unless p is linear will not in general be lifts.

2.4   Diffusion Operators with Projectible Symbols

Given $p : N \to M$ as before, suppose now that we have a diffusion operator $\mathcal{B}$ on M with a projectible symbol, c.f. Definition 2.1.5. This means that $\sigma^{\mathcal{B}}$ lies over some positive semi-definite linear map $\eta : T^*M \to TM$. *Assume that η has constant rank.* We will show that in this case we also have a decomposition of $\mathcal{B}$. To do this first choose some cohesive diffusion operator $\mathcal{A}$ on M with $\sigma^{\mathcal{A}} = \eta$. In general there is no canonical way to do this, though if η were non-degenerate we could choose $\mathcal{A}$ to be a multiple of the Laplace-Beltrami operator of the induced metric on M.

From above we also have an induced semi-connection with horizontal sub-bundle H, say, of TN.

Definition 2.4.1. We will say that $\mathcal{B}$ **descends cohesively** (over p) if it has a projectible symbol inducing a constant rank $\eta : T^*M \to TM$, and there exists a horizontal vector field, b^H, such that

$$\mathcal{B} - \mathbf{L}_{b^H}$$

is projectible over p.

The following is a useful observation. Its proof is immediate.

Proposition 2.4.2. *If $\mathcal{B}$ descends cohesively, then for each choice of $\mathcal{A}$ satisfying $\sigma^{\mathcal{A}}_{p(u)} = T_u p \sigma^{\mathcal{B}}_u (T_u p)^*$ there is a horizontal vector field b^H such that $\mathcal{B} - \mathbf{L}_{b^H}$ lies over $\mathcal{A}$.*

Lemma 2.4.3. *Assume that η has constant rank. If f is a function on M let $\tilde{f} = f \circ p$. For any choice of $\mathcal{A}$ with symbol η the map*

$$f \mapsto \mathcal{B}(\tilde{f}) - \widetilde{\mathcal{A}(f)}$$

is a derivation from $C^{\infty}M$ to $C^{\infty}N$ where any $f \in C^{\infty}M$ acts on $C^{\infty}N$ by multiplication by $\tilde{f}$.

Proof. The map is clearly linear and for smooth $f, g : M \to \mathbf{R}$ we have

$$\eta(\widetilde{df}, dg) = \sigma^{\mathcal{B}}(d\tilde{f}, d\tilde{g})$$

so by definition of symbols:

$$\mathcal{B}(\tilde{f}\tilde{g}) - \widetilde{\mathcal{A}(fg)} = \mathcal{B}(\tilde{f})\tilde{g} + \mathcal{B}(\tilde{g})\tilde{f} - \widetilde{\mathcal{A}(f)}\tilde{g} - \widetilde{\mathcal{A}(g)}\tilde{f}$$

as required. □

Let $\mathfrak{D}$ denote the space of derivations from $C^\infty M$ to $C^\infty N$ using the above action. Note that for $p^*TM \to N$ the pull-back of TM over p, the space $C^\infty \Gamma p^*TM$ of smooth sections of p^*TM can be considered as the space of smooth functions $V : N \to TM$ with $V(u) \in T_{p(u)}M$ for all $u \in N$. We can then define

$$\Theta : C^\infty \Gamma p^*TM \to \mathfrak{D}$$

by

$$\Theta(V)(f)(u) = df_{p(u)}(V(u)).$$

Lemma 2.4.4. *Assume that η has constant rank. The map $\Theta : C^\infty \Gamma p^*TM \to \mathfrak{D}$ is a linear bijection.*

Proof. Let $\eth \in \mathfrak{D}$. Fix $u \in N$. The map from $C^\infty M$ to $\mathbf{R}$ given by $f \mapsto \eth f(u)$ is a derivation at $p(u)$; here the action of any $f \in C^\infty M$ on $\mathbf{R}$ is multiplication by $f(p(u))$, and so corresponds to a tangent vector, $V(u)$ say, in $T_{p(u)}M$. Then $\eth f(u) = df_{p(u)}(V(u))$. By assumption $\eth f(u)$ is smooth in u, and so by suitable choices of f we see that V is smooth. Thus $\Theta(V) = \eth$ and Θ has an inverse. □

From these lemmas we see there exists $b \in C^\infty \Gamma p^*TM$ with the property that

$$\left(\mathcal{B}\tilde{f} - \widetilde{\mathcal{A}f}\right)(u) = df_{p(u)}(b(u)) \tag{2.28}$$

for all $u \in N$ and $f \in C^\infty M$. Assume that b has its image in the subbundle E of TM determined by η. Using the horizontal lift map h determined by $\mathcal{B}$, define a vector field b^H on N:

$$b^H(u) = h_u(b(u)).$$

Proposition 2.4.5. *Assume that η has constant rank and that b has its image in the subbundle E determined by η. The vector field b^H is such that $\mathcal{B} - b^H$ is over $\mathcal{A}$, and so $\mathcal{B}$ descends cohesively.*

Proof. For $f \in C^\infty M$,

$$(\mathcal{B} - b^H)(\tilde{f}) = \widetilde{\mathcal{A}f} + df(b(-)) - df \circ Tp(b^H(-)) = \widetilde{\mathcal{A}f}$$

using the fact that $Tp(b^H(-)) = b(-)$. □

We can now extend the decomposition theorem:

Theorem 2.4.6. *Let $\mathcal{B}$ be a diffusion operator on N which descends cohesively over $p : N \to M$. Then $\mathcal{B}$ has a unique decomposition:*

$$\mathcal{B} = \mathcal{B}^H + \mathcal{B}^V$$

into the sum of diffusion operators such that

(i) $\mathcal{B}^V$ *is vertical,*

(ii) $\mathcal{B}^H$ *is cohesive and $T_u p$ is injective on the image of $\sigma_u^{\mathcal{B}^H}$ for all $u \in N$.*

With respect to the induced semi-connection $\mathcal{B}^H$ is horizontal.

Proof. Using the notation of the previous proposition we know that $\mathcal{B} - b^H$ is over a cohesive diffusion operator $\mathcal{A}$. By Theorem 2.2.5 we have a canonical decomposition

$$\mathcal{B} - b^H = \mathcal{B}^1 + \mathcal{B}^V,$$

leading to

$$\mathcal{B} = (b^H + \mathcal{B}^1) + \mathcal{B}^V.$$

If we set $\mathcal{B}^H = b^H + \mathcal{B}^1$ we have a decomposition as required. On the other hand if we have two such decompositions of $\mathcal{B}$ we get two decompositions of $\mathcal{B} - b^H$. Both components of the latter must agree by the uniqueness in Theorem 2.2.5, and so we obtain uniqueness in our situation. $\square$

Extending Definition 2.2.3 we could say that a diffusion operator $\mathcal{B}^H$ satisfying condition (ii) in the theorem **has no vertical part**.

Note that if we drop the hypothesis that b^H is horizontal, or equivalently that b in Proposition 2.4.5 has its image in E, we still get a decomposition by taking an arbitrary lift of b to be b^H but we will no longer have uniqueness.

2.5   Horizontal lifts of paths and completeness of semi-connections

A semi-connection on $p : N \to M$ over a subbundle E of TM gives a procedure for horizontally lifting paths on M to paths on N as for ordinary connections but now we require the original path to have derivatives in E; such paths may be called **E-horizontal**.

Definition 2.5.1. A Lipschitz path $\tilde{\sigma}$ in N is said to be a **horizontal lift** of a path σ in M if

- $p \circ \tilde{\sigma} = \sigma$,

- the derivative of $\tilde{\sigma}$ almost surely takes values in the horizontal subbundle H of TN.

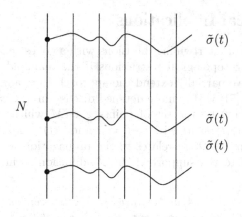

Note that a Lipschitz path $\sigma : [a, b] \to M$ with $\dot{\sigma}(t) \in \mathbf{E}_{\sigma(t)}$ for almost all $a \leqslant t \leqslant b$ has at most one horizontal lift from any starting point u_a in $p^{-1}(\sigma(a))$. To see this first note that any such lift must satisfy

$$\dot{\tilde{\sigma}}(t) = h_{\tilde{\sigma}(t)}\dot{\sigma}(t). \tag{2.29}$$

This equation can be extended to give an ordinary differential equation on all of N. For example take a smooth embedding $j : M \to \mathbf{R}^m$ into some Euclidean space. Set $\beta(t) = j(\sigma(t))$. Let $X(x) : \mathbf{R}^m \to E_x$ be the adjoint of the restriction of the derivative $T_x j$ of j to E_x, using some Riemannian metric on E. Then σ satisfies the differential equation

$$\dot{x}(t) = X(x(t))(\dot{\beta}(t)) \tag{2.30}$$

and it is easy to see that the horizontal lifts of σ are precisely the solutions of

$$\dot{u}(t) = h_{u(t)}X(p(u(t)))(\dot{\beta}(t))$$

starting from points above $\sigma(a)$ and lasting until time b.

In the generality in which we are working there may not be any such solutions, for example because of "holes" in N. We define the semi-connection to be **complete** if every Lipschitz path σ with derivatives in E almost surely has a horizontal lift starting from any point above the starting point of σ.

Note that completeness is assured if the fibres of N are compact, or if an X, with values in E, and β, can be found so that σ is a solution to equation (2.30) and there is a complete metric on N for which the horizontal lift of X is bounded on the inverse image of σ under p. In particular the latter will hold if p is a principal bundle and we have an equivariant semi-connection as in the next chapter. It will also hold if there is a complete metric on N for which the horizontal lift map $h_u \in \mathbb{L}(E_{p(u)}; T_u N)$ is uniformly bounded for u in the image of σ.

2.6   Topological Implications

Although our set-up of intertwining diffusions with a cohesive $\mathcal{A}$ seems quite general it implies strong topological restrictions if the manifolds are compact and more general. Here we partially extend the approach Hermann used for Riemannian submersions in [51] with a more detailed discussion in Chapter 6 below.

For this let $\mathcal{D}^0(x)$ be the set of points $z \in M$ which can be reached by Lipschitz curves $\sigma : [0, t] \to M$ with $\sigma(0) = x$ and $\sigma(t) = z$ with derivative in E almost surely. Its closure $\mathcal{D}'(x)$ relates to the propagation set for the maximum principle for $\mathcal{A}$, and to the support of the $\mathcal{A}$- diffusion as in Stroock-Varadhan [95], see Taira [99].

Theorem 2.6.1. *For $\mathcal{B}$ and $\mathcal{A}$ as before with $\mathcal{A}$ cohesive, take $x_0 \in M$ and $z \in \mathcal{D}^0(x_0)$. Assume the induced semi-connection is complete. Then if $p^{-1}(x_0)$ is a submanifold of N so is $p^{-1}(z)$ and they are diffeomorphic. Also if z is a regular value of p so is x.*

Proof. Let $\sigma : [0, T] \to M$ be a Lipschitz E-horizontal path from x to z. There is a smooth factorisation $\sigma_x^{\mathcal{A}} = X(x)X(x)^*$ for $X(x) \in \mathcal{L}(\mathbf{R}^m; T_x M)$, $x \in M$. Take the horizontal lift $\tilde{X} : \underline{\mathbf{R}}^m \to TN$ of X.

By the completeness hypothesis the time dependent ODE on N,

$$\frac{dy_s}{ds} = \tilde{X}(y_s)X\left(\sigma(s)|_{[\ker X(x_0)]^\perp}\right)^{-1}(\dot{\sigma}(s))$$

will have solutions from each point above $\sigma(0)$ defined up to time T and so a flow giving the required diffeomorphism of fibres. Moreover, by the usual lower semi-continuity property of the "explosion time", this holonomy flow gives a diffeomorphism of a neighbourhood of $p^{-1}(x)$ in N with a neighbourhood of the fibre above z. The diffeomorphism commutes with p. Thus if one of x and z is a regular value so is the other. $\square$

Corollary 2.6.2. *Assume the conditions of the theorem and that E satisfies the standard Hörmander condition that the Lie algebra of vector fields generated by sections of E spans each tangent space $T_y M$ after evaluation at y. Then p is a submersion all of whose fibres are diffeomorphic.*

Proof. The Hörmander condition implies that $\mathcal{D}^0(x) = M$ for all $x \in M$ by Chow's theorem (*e.g.* see Sussmann [98] or [49]. In [49] Gromov shows that under this condition any two points of M can be joined by a smooth E-horizontal curve. $\square$

Corollary 2.6.3. *Assume the conditions of the theorem and that $\mathcal{D}^0(x)$ is dense in M for all $x \in M$ and $p : N \to M$ is proper. Then p is a locally trivial bundle over M.*

Proof. Take $x \in M$. The set $Reg(p)$ of regular values of p is open by our properness assumption. It is also non-empty, even dense in M, by Sard's theorem, and so since $\mathcal{D}^0(x)$ is dense, there exists a regular value z which is in $\mathcal{D}^0(x)$. It follows from the theorem that $x \in Reg(p)$, and so p is a submersion. However it is a well-known consequence of the inverse function theorem that a proper submersion is a locally trivial bundle.   □

Note that we only need $Reg(p)$ to be open, rather than p proper, to ensure that p is a submersion. The density of $\mathcal{D}^0(x)$ can hold because of global behaviour, for example if M is a torus and E is tangent to the foliation given by an irrational flow.

2.7 Notes

Intertwined stochastic differential equations

Consider p-related stochastic differential equations as in Section 2.3.1:

$$du_t = \tilde{X}(u_t) \circ dB_t + \tilde{A}(u_t) \, dt,$$
$$dx_t = X(x_t) \circ dB_t + A(x_t) \, dt.$$

Then not only are their generators $\mathcal{B}$ and $\mathcal{A}$ interwined but also the operators $\mathcal{B}^\wedge$ and $\mathcal{A}^\wedge$, which they induce on differential forms, and their semigroups $\tilde{P}^\wedge$ and $P_\cdot^\wedge$:

$$\mathcal{B}^\wedge \circ p^* = p^* \circ \mathcal{A}^\wedge \qquad \text{and} \qquad \tilde{P}^\wedge \circ p^* = p^* \circ P^\wedge.$$

Recall that the pull-back $f^*\phi$ of a differential form ϕ by a differentiable map $f : N \to M$ is given by $f^*\phi(v^1, \ldots, v^k) = \phi_{f(u)}(T_u f(v^1), \ldots, T_u f(v^k))$ for $v^1, \ldots, v^k$ in $T_u N$ when ϕ is a k-form on M. The operators are given by the same formulae as for functions:

$$\mathcal{A}^\wedge = \frac{1}{2} \sum_{j=1}^m \mathbf{L}_{X^j} \mathbf{L}_{X^j} + \mathbf{L}_A, \tag{2.31}$$

where now the Lie derivatives are acting on forms. Thus, for example,

$$\mathbf{L}_A \phi = \frac{d}{dt} \left((\eta_t^A)^* \phi \right) |_{t=0} \tag{2.32}$$

where η^A denotes the flow of the vector field A. The semi-groups are given by:

$$P_t^\wedge \phi = \mathbf{E}(\xi_t)^* \phi \tag{2.33}$$

when the integrals exist, with modifications if the flow is only locally defined. There is a detailed discussion in [36].

That the operators are intertwined is clear from equation (2.32) and the fact that the flows of the vector fields involved are intertwined by p. Similarly the intertwining of the semi-groups comes from the fact that the flows of the SDE's are intertwined.

In fact there is the stronger result, following in the same way, that the "co-differentials" $\hat{\delta}$ and $\hat{\delta}$ determined by the two SDE's are intertwined. These take k-forms to $(k-1)$-forms and are defined by:

$$\hat{\delta} = -\sum_{j=1}^{m} \iota_{X^j} \mathbf{L}_{X^j} \qquad (2.34)$$

where ι denotes interior product. It is shown in [36], page 37, that

$$\mathcal{A}^{\wedge} = -\frac{1}{2}\left(\hat{\delta}d + d\hat{\delta}\right) + \mathbf{L}_A$$

where d is the usual exterior derivative.

There is further discussion about this and its relationships with the more classical results in [71], [105], [106], and [48] in the Notes to Chapter 7, and the Appendix, Section 9.5.2.

Chapter 3

Equivariant Diffusions on Principal Bundles

Let M be a smooth finite dimensional manifold and $P(M,G)$ a principal fibre bundle over M with structure group G a Lie group. Denote by $\pi : P \to M$ the projection and R_a right translation by a. Consider on P a diffusion generator $\mathcal{B}$, which is **equivariant**, i.e. for all $f \in C^2(P; \mathbf{R})$,

$$\mathcal{B}f \circ R_a = \mathcal{B}(f \circ R_a), \qquad a \in G.$$

Set $f^a(u) = f(ua)$. Then the above equality can be written as $\mathcal{B}f^a = (\mathcal{B}f)^a$. The operator $\mathcal{B}$ induces an operator $\mathcal{A}$ on the base manifold M. Set

$$\mathcal{A}f(x) = \mathcal{B}(f \circ \pi)(u), \qquad u \in \pi^{-1}(x), f \in C^2(M), \tag{3.1}$$

which is well defined since

$$\mathcal{B}(f \circ \pi)(u \cdot a) = \mathcal{B}((f \circ \pi)^a)(u) = \mathcal{B}((f \circ \pi))(u).$$

Example 3.0.1. One of the simplest examples is obtained from the map $p : SO(n+1) \to S^n$ defined by choosing some point x_0 on S^n, considering the natural action of $SO(n+1)$ on the sphere by rotation, and setting $p(u) = u.x_0$ for $u \in SO(n+1)$. This has the natural structure of a principal bundle with group $SO(n)$ when we identify $SO(n)$ with the subgroup of $SO(n+1)$ which fixes x_0. If we take the bi-invariant Riemannian metric on $SO(n+1)$ determined by the Hilbert-Schmidt inner product $\langle A, B \rangle_{HS} := \text{trace}(B^*A)$ on the Lie algebra $\mathfrak{so}(n+1)$, identified with the space of skew-symmetric $(n+1) \times (n+1)$-matrices, the projection p is a Riemannian submersion, see Chapter 7 below, but onto the sphere of radius $\sqrt{2}$. It therefore sends the Laplacian of $SO(3)$, which is bi-invariant under the full group $SO(3)$, onto the Laplacian of $S^n(\sqrt{2})$.

K.D. Elworthy et al., *The Geometry of Filtering*, Frontiers in Mathematics, DOI 10.1007/978-3-0346-0176-4_3, © Springer Basel AG 2010

Let $E^{n+1}_{[p,q]}$ be the elementary $(n+1) \times (n+1)$- matrix with entries which are all zero except for that in the p-th row and q-th column which is 1. Set $A^{n+1}_{[p,q]} = \frac{1}{\sqrt{2}}(E^{n+1}_{[p,q]} - E^{n+1}_{[p,q]})$. Then $\{A^{n+1}_{[p,q]}\}_{1 \leqslant p < q \leqslant n+1}$ forms an orthonormal base for $\mathfrak{so}(n+1)$ with this inner product. If we let $A^*_{[p,q]}$ denote the left invariant vector field on $SO(n+1)$ determined by $A^{n+1}_{[p,q]}$, our Laplacian of $SO(n+1)$ has the sum of squares representation

$$\triangle^{SO(n)} = \frac{1}{2} \sum_{1 \leqslant p < q \leqslant n+1} \sqrt{2}A^*_{[p,q]} \sqrt{2}A^*_{[p,q]}.$$

If we want $\mathcal{A}$ to be the usual Laplacian for S^n, rather than for $S^n(\sqrt{2})$, we can modify the metric of $SO(3)$ to be $\frac{1}{2}$ times that given by the Hilbert-Schmidt norm. Denoting the resulting Laplacian by $\mathcal{B}$ we then have

$$\mathcal{B} = \frac{1}{2} \sum_{1 \leqslant p < q \leqslant n+1} 2A^*_{[p,q]} 2A^*_{[p,q]}.$$

3.1 Invariant Semi-connections on Principal Bundles

Definition 3.1.1. Let E be a subbundle of TM and $\pi : P \to M$ a principal G-bundle. An **invariant semi-connection** over E, or **principal semi-connection** in the terminology of Michor, on $\pi : P \to M$ is a smooth subbundle $H^E TP$ of TP such that

(i) $T_u\pi$ maps the fibres $H^E T_uP$ bijectively onto $E_{\pi(u)}$ for all $u \in P$.

(ii) $H^E TP$ is G-invariant.

Notes.

1. Such a semi-connection determines and is determined by, a smooth horizontal lift:

$$h_u : E_{\pi(u)} \to T_uP$$

   such that

   (i) $T_u\pi \circ h_u(v) = v$, for all $v \in E_x \subset T_xM$;

   (ii) $h_{u \cdot a} = T_u R_a \circ h_u$.

2. The action of G on P induces a homomorphism of the Lie algebra $\mathfrak{g}$ of G with the algebra of left invariant vector fields on P: if $A \in \mathfrak{g}$,

$$A^*(u) = \frac{d}{dt}\bigg|_{t=0} u \exp(tA), \qquad u \in P,$$

   and A^* is called the fundamental vector field corresponding to A. Note that for $a \in G$:

$$A^*(ua) = \text{ad}(a)A^*(u) \qquad (3.2)$$

for $\mathrm{ad}(a) : \mathfrak{g} \to \mathfrak{g}$ the adjoint action of a.

Using the splitting (2.5) of F_u our semi-connection determines, (and is determined by), a 'semi-connection one-form' $\varpi \in \mathcal{L}(H + VTP; \mathfrak{g})$ which vanishes on H has $\varpi(A^*(u)) = A$, and

$$\varpi_{ua}TR_a(V) = \mathrm{ad}(a^{-1})\varpi_u(V) \qquad V \in H_u + VT_uP. \tag{3.3}$$

3. Let F be an associated vector bundle to P with fibre V. An E-semi-connection on P gives a covariant derivative $\nabla_w Z \in F_x$ for $w \in E_x$, $x \in M$ where Z is a section of F. This is defined, as usual for connections, by

$$\nabla_w Z = u\big(d(\tilde{Z})(h_u(w))\big),$$

$u \in \pi^{-1}(x)$. Here $\tilde{Z} : P \to V$ is

$$\tilde{Z}(u) = u^{-1}Z(\pi(u))$$

considering u as an isomorphism $u : V \to F_{\pi(u)}$. This agrees with the 'semi-connections on E' defined in Elworthy-LeJan-Li, [36], when P is taken to be the linear frame bundle of TM and $F = TM$. Such semi-connections could be called 'linear semi-connections' since they arise from a principal semi-connection via a linear representation, but since we will only deal with such connections when the vector bundle structure is being used we shall usually take as read the qualification 'linear'.

Theorem 3.1.2. *Assume σ^A has constant rank. Then σ^B gives rise to an invariant semi-connection on the principal bundle P whose horizontal map is given by (2.4).*

Proof. It has been shown that h_u is well defined by (2.4). Next we show h_u defines a semi-connection. As noted earlier, h defines a semi-connection if (i) $T_u\pi\circ h_u(v) = v$, $v \in E_x \subset T_xM$ and (ii) $h_{u\cdot a} = T_uR_a \circ h_u$. The first is immediate by Lemma 2.1.1 and for the second observe $\pi \circ R_a = \pi$. So $T\pi \circ TR_a = T\pi$ and $(T\pi)^* = (TR_a)^* \cdot (T\pi)^*$ while the following diagram

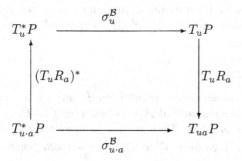

commutes by equivariance of $\mathcal{B}$. Therefore

$$
\begin{aligned}
T_u R_a \circ h_u &= T_u R_a \cdot \sigma_u^{\mathcal{B}} \left(T_u \pi\right)^* \circ \left(\sigma_x^{\mathcal{A}}\right)^{-1} \\
&= T_u R_a \cdot \sigma_u^{\mathcal{B}} \circ \left(T_u R_a\right)^* \circ \left(T_{u \cdot a} \pi\right)^* \circ \left(\sigma_x^{\mathcal{A}}\right)^{-1} \\
&= \sigma_{u \cdot a}^{\mathcal{B}} \circ \left(T_{u \cdot a} \pi\right)^* \circ \left(\sigma_x^{\mathcal{A}}\right)^{-1} = h_{u \cdot a}. \qquad \square
\end{aligned}
$$

Curvature forms and holonomy groups etc for semi-connections are defined analogously to those associated two connections, we note the following:

Proposition 3.1.3. *In the situation of Proposition 2.1.4, suppose $\mathcal{A}$ is elliptic, p^1, p^2 are principal bundles with groups G^1 and G^2 respectively, and F is a homomorphism of principal bundles with corresponding homomorphism $f : G^1 \to G^2$. Let Γ^1 and Γ^2 be the semi-connections on N^1, N^2 determined by $\mathcal{B}^1$ and $\mathcal{B}^2$. Then:*

(i) *Γ^2 is the unique semi-connection on $p^2 : N^2 \to M$ such that TF maps the horizontal subspaces of TN^1 into those of TN^2.*

(ii) *If ω^j, Ω^j are the semi-connection and curvature form of Γ^j, for $j = 1, 2$, then*

$$F^*(\omega^2) = f_* \circ \omega^1$$

and

$$F^*(\Omega^2) = f_* \circ \Omega^1$$

for $f_* : \underline{g}_1 \to \underline{g}_2$ the homomorphism of Lie algebras induced by f.

(iii) *Moreover $f : G^1 \to G^2$ maps the Γ^1 holonomy group at $u \in N^1$ onto the Γ^2 holonomy group at $F(u)$ for each $u \in N^1$ and similarly for the restricted holonomy groups.*

Proof. Proposition 2.1.4 assures us that TF maps horizontal to horizontal. Uniqueness together with (ii), (iii) come as in Kobayashi-Nomizu [56] (Proposition 6.1 on p79). $\qquad \square$

3.2   Decompositions of Equivariant Operators

Take a basis $A_1, \ldots, A_n$ of $\mathfrak{g}$ with corresponding fundamental vector fields $\{A_i^*\}$. Write the semi-connection 1-form as $\varpi = \sum \varpi^k A_k$ so that ϖ^k are real-valued, partially defined, 1-forms on P.

In our equivariant situation we can give a more detailed description of the decomposition in Theorem 2.2.5.

Theorem 3.2.1. *Let $\mathcal{B}$ be an equivariant operator on P and $\mathcal{A}$ be the induced operator on the base manifold. Assume that $\mathcal{A}$ is cohesive and let $\mathcal{B} = \mathcal{A}^H + \mathcal{B}^V$ be the decomposition of Theorem 2.2.5. Then $\mathcal{B}^V$ has a unique expression of the form $\sum \alpha^{ij} \mathcal{L}_{A_i^*} \mathcal{L}_{A_j^*} + \sum \beta^k \mathcal{L}_{A_k^*}$, where α^{ij} and β^k are smooth functions on P,* *given by* $\alpha^{k\ell} = \varpi^k \left(\sigma^{\mathcal{B}}(\varpi^\ell) \right)$, *and* $\beta^\ell = \delta^{\mathcal{B}}(\varpi^\ell)$ *for* ϖ *the semi-connection 1-form on* P. *Define* $\alpha : P \to \mathfrak{g} \otimes \mathfrak{g}$ *and* $\beta : P \to \mathfrak{g}$ *by*

$$\alpha(u) = \sum \alpha^{ij}(u) A_i \otimes A_j, \qquad \beta(u) = \sum \beta^k(u) A_k. \qquad (3.4)$$

These are independent of the choices of basis of $\mathfrak{g}$ *and are equivariant:*

$$\alpha(ug) = (\mathrm{ad}(g) \otimes \mathrm{ad}(g))\, \alpha(u)$$

and

$$\beta(ug) = \mathrm{ad}(g)\beta(u).$$

Proof. Since every vertical vector field is a linear combination of the fundamental vertical vector fields, Proposition 1.4.7, shows that

$$\mathcal{B}^V = \sum \alpha^{i,j} \mathcal{L}_{A_i^*} \mathcal{L}_{A_j^*} + \sum \beta^k \mathcal{L}_{A_k^*}$$

for certain functions α^{ij}, β^k. For $f, g : P \to \mathbf{R}$ setting $\sigma := \sigma^{\mathcal{B} - \mathcal{A}^H}$,

$$\begin{aligned}
df\left(\sigma(dg)\right) &= \frac{1}{2} \sum \alpha^{i,j} \mathcal{L}_{A_i^*} \mathcal{L}_{A_j^*}(fg) - \frac{1}{2} \sum g\alpha^{i,j} \mathcal{L}_{A_i^*} \mathcal{L}_{A_j^*}(f) \\
&\quad - \frac{1}{2} \sum f\alpha^{i,j} \mathcal{L}_{A_i^*} \mathcal{L}_{A_j^*}(g) \\
&= \sum \alpha^{i,j} \mathcal{L}_{A_i^*}(f) \mathcal{L}_{A_j^*}(g) \\
&= \sum \alpha^{i,j} df(A_i^*) dg(A_j^*).
\end{aligned}$$

Since $\varpi(A_k^*) = A_k$, we see that $\varpi^k(A_\ell^*) = \delta_{k\ell}$ and

$$\varpi^k(\sigma(\varpi^\ell)) = \sum \alpha^{i,j} \delta_{ik} \delta_{j\ell} = \alpha^{k\ell}.$$

Since $\mathcal{A}^H$ is horizontal $\sigma^{\mathcal{A}^H}$ has its image in the horizontal tangent bundle and so is annihilated by ϖ^k. Thus

$$\alpha^{k\ell} = \varpi^k\left(\sigma^{\mathcal{B}}(\varpi^\ell)\right). \qquad (3.5)$$

Note that by the characterisation, Proposition 1.2.1,

$$\delta^{\mathcal{B}^V} = \sum \alpha^{i,j} \mathcal{L}_{A_i^*} \iota_{A_j^*} + \sum \beta^k \iota_{A_k^*}.$$

Since $\varpi^\ell(A^{*\ell})$ is identically 1, it follows that $\delta^{\mathcal{B}^V}(\varpi) = \beta^\ell$. Again $\delta^{\mathcal{A}^H}(\varpi^\ell) = 0$ and so

$$\beta^\ell = \delta^{\mathcal{B}}(\varpi^\ell) \qquad (3.6)$$

as required.

For the last part α and β can be considered as obtained from the extension of the symbol $\sigma^{\mathcal{B}}$ and $\delta^{\mathcal{B}}$ to $\mathfrak{g}$-valued two- and one-forms respectively: $\alpha = \varpi(-)\sigma^{\mathcal{B}}\varpi(-)$ and $\beta = \delta^{\mathcal{B}}(\varpi(-))$. To make this precise consider $\sigma_u^{\mathcal{B}}$ as a bilinear form and so as a linear map

$$\sigma_u^{\mathcal{B}} : T_u^*P \otimes T_u^*P \to \mathbf{R}.$$

The extension is the trivial one given by

$$\sigma_u^{\mathcal{B}} \otimes 1 \otimes 1 : T_u^*P \otimes T_u^*P \otimes \mathfrak{g} \otimes \mathfrak{g} \to \mathbf{R} \otimes \mathfrak{g} \otimes \mathfrak{g} \simeq \mathfrak{g} \otimes \mathfrak{g}$$

using the identification of $T_u^*P \otimes \mathfrak{g}$ with $L(T_uP; \mathfrak{g})$. Similarly the extension of $\delta^{\mathcal{B}}$ is

$$\delta_u^{\mathcal{B}} \otimes 1 : T_u^*P \otimes \mathfrak{g} \to \mathbf{R} \otimes \mathfrak{g} \simeq \mathfrak{g}.$$

Thus

$$\alpha(u)(\omega \otimes \omega) = \left(\sigma^{\mathcal{B}} \otimes 1 \otimes 1\right)(P_{23}\omega \otimes \omega)$$

where $P_{23} : T^*P \otimes \mathfrak{g} \otimes T^*P \otimes \mathfrak{g} \to T^*P \otimes T^*P \otimes \mathfrak{g} \otimes \mathfrak{g}$ is the standard permutation and $\beta_u(\omega) = (\delta_u^{\mathcal{B}} \otimes 1)(\omega)$.

The equivariance of ϖ,

$$(R_g)^*\varpi = ad(g^{-1})(\varpi), \qquad g \in G,$$

is equivalent to the invariance of ϖ when considered as a section of $T^*M \otimes \mathfrak{g}$ under

$$TR_g \otimes ad(g) : T^*M \otimes \mathfrak{g} \to T^*M \otimes \mathfrak{g}, \qquad g \in G. \qquad \square$$

Remark 3.2.2. (a) For any equivariant operator of the form $\mathcal{B} = \sum_{i,j} \alpha^{ij} L_{A_i^*} L_{A_j^*}$ $+ \sum \beta^k L_{A_k^*}$ with $(\alpha^{ij}(u))$ positive semi-definite for each $u \in P$ we can define maps α and β by (3.4). Note that $\alpha(u)$ is essentially the symbol of $\mathcal{B}$ restricted to the fibre $P_{\pi(u)}$ through u:

$$\sigma_u^{\mathcal{B}}|_{P_{\pi(u)}} : T_u^*P_{\pi(u)} \to T_uP_{\pi(u)}$$

with ϖ_u identifying $T_uP_{\pi(u)}$ with $\mathfrak{g}$. Similarly β determines $\delta^{\mathcal{B}}$ on a basis of sections of $(VTP)^*$.

(b) Let $\{u_t : 0 \leqslant t \leqslant \varsigma\}$ be a $\mathcal{B}$-diffusion on P. By (3.5), $2\alpha^{kl}(u_t)$ is the derivative of the bracket $\left\langle \int_0^{\cdot} \varpi_{u_s}^k \circ du_s, \int_0^{\cdot} \varpi_{u_s}^l \circ du_s \right\rangle$ of the integrals of ω^k and ω^l along $\{u_t : 0 \leqslant t < \varsigma\}$. See Chapter 4 below for a detailed discussion. Thus $\alpha(u_t)$ is the derivative of the tensor quadratic variation:

$$\alpha(u_t) = \frac{1}{2}\frac{d}{dt}\int_0^t \left(\varpi_{u_t} \circ du_t \otimes \varpi_{u_t} \circ du_t\right).$$

Moreover by (3.6) and Lemma 4.1.2 below $\int_0^t \beta(u_s)ds$ is the bounded variation part of $\int_0^t \varpi_{u_s} \circ du_s$.

(c) If we fix $u_0 \in P$ and take an inner product on $\mathfrak{g}$ we can diagonalise $\alpha(u_0)$ to write

$$\alpha(u_0) = \sum_n \mu_n A_n \otimes A_n$$

where $\{A_n : n = 1, \ldots \dim(\mathfrak{g})\}$ is an orthonormal basis. The μ_n are the eigenvalues of $\alpha(u_0)^{\#} : \mathfrak{g} \to \mathfrak{g}$ obtained using the isomorphism:

$$\mathfrak{g} \otimes \mathfrak{g} \quad \to \quad \mathcal{L}(\mathfrak{g}; \mathfrak{g})$$
$$a \otimes b \quad \mapsto \quad (a \otimes b)^{\#},$$

where $(a \otimes b)^{\#}(v) = \langle b, v \rangle a$.

Note that for $g \in G$, $\alpha(u_0 \cdot g) = \sum_n \mu_n \operatorname{ad}(g) A_n \otimes \operatorname{ad}(g) A_n$. When the inner product is $\operatorname{ad}(G)$-invariant, then $\{\operatorname{ad}(g) A_n\}_{n=1}^{\dim(\mathfrak{g})}$ is still orthonormal and the $\{\mu_n\}_n$ are the eigenvalues of $\alpha(u_0 \cdot g)^{\#}$. They are therefore independent of the choice of u_0 in a given fibre, (but depend on the inner product chosen).

Example 3.2.3. Using the notation of Example 2.2.11, let P be the special orthonormal frame bundle of the two-dimensional horizontal distribution of the Heisenberg group H, whose fibre at $(x, y, z) \in H$ are orthogonal frames with values in $E = \operatorname{span}\{X, Y\}$. Note that E, and so P, is trivialised by the left action of H and using this we shall consider P as the product $H \times SO(2)$ with projection $(x, y, z, A) \mapsto (x, y, z)$. The actual frames are compositions of the rotation A on (u, v) and the map $(u, v) \mapsto (u, v, \frac{1}{2}(xu - yv))$. Identifying $SO(2)$ with the circle S^1 the bundle P becomes a principal bundle with group S^1 acting on the right to be written as: $\pi : H \times S^1 \to H$. Then the Lie algebra $\mathfrak{g} = \mathfrak{s}^1$, which we shall identify with the imaginary axis $i\mathbf{R}$. The tangent space to S^1 at a general point $e^{i\theta}$ of S^1 will then be represented by $\{ie^{i\theta} a : a \in \mathbf{R}\} \subset \mathbb{C}$.

An invariant semi-connection on P may be defined by its semi-connection 1-form ϖ, which here will be a section of the dual bundle to the bundle $E \times TS^1 \to H \times S^1$, but with values in $i\mathbf{R}$. In our situation this corresponds to a horizontal vector field $V_0 = v_1 X + v_2 Y$. (As a vector potential determines the connection in the gauge-theoretic interpretation of electromagnetism.) The vector field defines such an ϖ given on $E \times TS^1$ by

$$\varpi_{(x,y,z,e^{i\theta})}(w, a) = e^{-i\theta} a + i\langle w, V_0(x, y, z)\rangle_E.$$

The horizontal subspace HTP for this connection is given at $(x, y, z, e^{i\theta})$ by:

$$H_{(x,y,z,e^{i\theta})}TP = \{(w, a) : e^{-i\theta} a + i\langle w, V_0(x, y, z)\rangle_E = 0\}.$$

Consider a path σ on H_3 with $\dot\sigma \in E$. Denote its horizontal lift starting at $(\sigma(0), 1)$ by $\tilde\sigma(t) = (\sigma(t), e^{i\theta(t)})$, some $\theta(t)$. Then

$$0 = \varpi(\dot{\tilde\sigma}_t) = i\dot\theta(t) + i\langle V_0(\sigma(t)), \dot\sigma(t)\rangle.$$

giving
$$\tilde{\sigma}(t) = (\sigma(t), e^{-i\int_0^t \langle V(\sigma_s), \dot{\sigma}(s)\rangle ds}).$$

Note that an E-semi-connection on the principal bundle E allows one to covariantly diffferentiate sections of E in E-directions; see Note 3 at the beginning of Section 3.1. The covariant derivative of such a section U is given by

$$\hat{\nabla}_w U = \nabla_w^L U + i\langle V_0(x,y,z), w\rangle_E U(x,y,z), \qquad w \in E_{(x,y,z)} \qquad (3.7)$$

where ∇^L refers to the left invariant connection on our Lie group H and we are treating E as a complex line bundle via its trivialisation.

In fact this shows that our semi-connection on E is the restriction of an E-semi-connection on the full tangent bundle TH of H since ∇^L is defined on all C^1 vector fields on H and $E \subset H$ so the right-hand side of the formula (3.7) is defined as an element of TH for all C^1 vector fields on H and $w \in E$. As for standard connections, semi-connections are determined by suitable covariant differentiation operators.

Next consider $\mathcal{A} = \frac{1}{2}(X^2 + Y^2)$ and for real-valued functions r_1, r_2, γ on H with $\gamma \geqslant r_1^2 + r_2^2$ set

$$\mathcal{B} = \frac{1}{2}\left\{(X^2 + Y^2) + r_1 \frac{\partial^2}{\partial x \partial \theta} + r_2 \frac{\partial^2}{\partial y \partial \theta} + \gamma \frac{\partial^2}{\partial^2 \theta} + (xr_2 - r_1 y)\frac{\partial^2}{\partial z \partial \theta} + (\frac{\partial r_1}{\partial x} + \frac{\partial r_2}{\partial y})\frac{\partial}{\partial \theta}\right\}.$$

Now we are treating $\theta \mapsto e^{i\theta}$ as giving charts for S^1 and $\frac{\partial}{\partial \theta}$ refers to the use of these local co-ordinates. Thus $\frac{\partial}{\partial \theta}$ corresponds to the tangent vector whose value at $e^{i\theta}$ is $ie^{i\theta}$ in our representation in $\mathbb{C}$. We shall find the vector field V_0 corresponding to the semi-connection induced by $\mathcal{B}$. If we complete the squares for $\mathcal{B}$, recalling that X and Y are given by $\frac{\partial}{\partial x} - \frac{1}{2}y\frac{\partial}{\partial z}$ and $\frac{\partial}{\partial y} + \frac{1}{2}x\frac{\partial}{\partial z}$ respectively, we obtain:

$$\mathcal{B} = \frac{1}{2}\left(\tilde{X}^2 + \tilde{Y}^2\right) + \frac{1}{2}\left((\gamma - r_1^2 - r_2^2)\frac{\partial^2}{\partial^2 \theta} - \frac{1}{2}(x\frac{\partial r_2}{\partial z} - y\frac{\partial r_1}{\partial z})\frac{\partial}{\partial \theta}\right) \qquad (3.8)$$

where $\tilde{X} = X + r_1 \frac{\partial}{\partial \theta}$ and $\tilde{Y} = Y + r_2 \frac{\partial}{\partial \theta}$. As vector fields on $H \times S^1$ this gives $\tilde{X} = (X, ir_1 e^{-i\theta})$ and $\tilde{Y} = (Y, ir_2 e^{i\theta})$.

Now $\frac{1}{2}(\tilde{X}^2 + \tilde{Y}^2)$ is cohesive and $T_u\pi$ is injective on the image of its symbol, which is $\{(a, b, i(ar_1 + br_2)e^{i\theta}) : a, b, \in \mathbf{R}\} \subset E_{(x,y,z)} \times T_{e^{i\theta}}S^1$ at $u = (x, y, z, e^{i\theta})$, and so, by Proposition 2.2.2, we see $\frac{1}{2}(\tilde{X}^2 + \tilde{Y}^2)$ is the lift of $\mathcal{A}$. The second term in equation (3.8) is clearly vertical hence (3.8) is the decomposition for the semi-connection determined by $(\mathcal{A}, \mathcal{B})$, using Theorem 2.4.6. That is:

$$\mathcal{A}^H = \frac{1}{2}(\tilde{X}^2 + \tilde{Y}^2) \quad \text{and} \quad \mathcal{B}^V = \frac{1}{2}\left((\gamma - r_1^2 - r_2^2)\frac{\partial^2}{\partial^2 \theta} - \frac{1}{2}(x\frac{\partial r_2}{\partial z} - y\frac{\partial r_1}{\partial z})\frac{\partial}{\partial \theta}\right).$$

We see that $\tilde{X} := h_u(X), \tilde{Y} := (h_u Y)$ and the horizontal lift map is:

$$h_{(x,y,z,e^{i\theta})}(a, b, \frac{1}{2}(xb - ya)) = (a, b, \frac{1}{2}(xb - ya), i(r_1 a + r_2 b)e^{i\theta}).$$

This horizontal lifting map is determined by the vector field $V_0 := -r_1 X - r_2 Y$. In particular a couple of choices of r_1 and r_2 give some interesting semi-connections. Letting $r_1 = -y$, $r_2 = x$, we have

$$\mathcal{B}_1 := \frac{1}{2}\left\{ (X^2 + Y^2) - y\frac{\partial^2}{\partial x \partial \theta} + x\frac{\partial^2}{\partial y \partial \theta} + \gamma\frac{\partial^2}{\partial^2 \theta} + (x^2 + y^2)\frac{\partial^2}{\partial z \partial \theta} \right\}$$

which determines the semi- connection

$$\left(a, b, \frac{1}{2}(xb - ya)\right) \mapsto \left(a, b, \frac{1}{2}(xb - ya), i(-ya + xb)e^{i\theta}\right).$$

Taking $r_1 = x, r_2 = y$, we have

$$\mathcal{B}_2 = \frac{1}{2}\left\{ (X^2 + Y^2) + x\frac{\partial^2}{\partial x \partial \theta} + y\frac{\partial^2}{\partial y \partial \theta} + \gamma\frac{\partial^2}{\partial^2 \theta} + 2\frac{\partial}{\partial \theta} \right\}$$

with semi-connection $(a, b, \frac{1}{2}(xb - ya)) \mapsto (a, b, \frac{1}{2}(xb - ya), i(xa + yb)e^{i\theta})$.

We return to this example in Example 3.4.3 below.

3.3 Derivative Flows and Adjoint Connections

Let $\mathcal{A}$ on M be given in Hörmander form

$$\mathcal{A} = \frac{1}{2}\sum_{j=1}^{m} \mathcal{L}_{X^j}\mathcal{L}_{X^j} + \mathcal{L}_A \tag{3.9}$$

for some smooth vector fields $X^1, \ldots X^m$, A. As before let E_x be the linear span of $\{X^1(x), \ldots, X^m(x)\}$ and assume $\dim E_x$ is constant, denoted by p, giving a subbundle $E \subset TM$. The vector fields $\{X^1(x), \ldots, X^m(x)\}$ determine a vector bundle map

$$X : \underline{\mathbf{R}}^m \to TM$$

with $\sigma^{\mathcal{A}} = X(x)X(x)^*$.

We can, and will, consider X as a map $X : \underline{\mathbf{R}}^m \to E$. Let Y_x be the right inverse $[X(x)|_{\ker X(x)^\perp}]^{-1}$ of $X(x)$ and $\langle , \rangle_x$ the inner product, induced on E_x by Y_x. Then X projects the flat connection on $\mathbf{R}^m$ to a metric connection $\check{\nabla}$ on E defined by

$$\check{\nabla}_v U = X(x)d[y \mapsto Y_y U(y)](v), \qquad U \in C^1\Gamma E, v \in T_y M. \tag{3.10}$$

(In [36] we have studied the properties of this construction together with the SDE induced by X, and there $\check{\nabla}$ is referred as the LW connection for the SDE; see also Section 9.5.2 in the Appendix.) Moreover any connection ∇ on a subbundle E of TM has an adjoint semi-connection ∇' on TM over E defined by

$$\nabla'_U V = \nabla_V U + [U, V], \qquad U \in \Gamma E, V \in \Gamma TM.$$

Remark 3.3.1. Note the converse of this is not true. In the discussion of Example 3.2.3 we noted shortly after equation (3.7) that our construction determined a semi-connection over E on the the whole tangent bundle. This could not be the adjoint of any connection on E because brackets of E-valued vector fields are not E-valued in general. In fact it is noted in [36] that parallel translation using the adjoint of a connection on a subbundle E of a tangent bundle can only preserve E if E is integrable.

Let $\pi : GLM \to M$ be the frame bundle of M, so $u \in \pi^{-1}(x)$ is a linear isomorphism $u : \mathbf{R}^n \to T_x M$. It is a principal bundle with group $GL(n)$. If $g \in GL(n)$ and $\pi(u) = x$, then $u \cdot g : \mathbf{R}^n \to T_x M$ is just the composition of u with g.

Any smooth vector field A on M determines smooth vector fields A^{TM} and A^{GL} on TM and GLM respectively as follows: Let $\eta_t : t \in (-\epsilon, \epsilon)$ be a (partial) flow for A and $T\eta_t$ its derivative. Then $v \mapsto T\eta_t(v)$ is a partial flow on TM and $u \mapsto T\eta_t \circ u$ one on GLM, Let A^{TM} and A^{GL} be the vector fields generating these flows. In fact A^{TM} is $\tau \circ TA : TM \to TTM$ where $\tau : TTM \to TTM$ is the canonical twisting map:

$$\tau(x, v, w, v') = (x, v, v', w)$$

in local coordinates.

Using this, the choice of our Hörmander form representation induces a diffusion operator $\mathcal{B}$ on GLM by setting

$$\mathcal{B} = \frac{1}{2} \sum \mathbf{L}_{(X^j)^{GL}} \mathbf{L}_{(X^j)^{GL}} + \mathbf{L}_{A^{GL}}.$$

This is invariant under the action of $GL(n)$ and π intertwines $\mathcal{B}$ and $\mathcal{A}$. For cohesive $\mathcal{A}$, a principal semi-connection is therefore induced on GLM. In turn this induces a semi-connection on TM with covariant derivative ∇, say, as described in Note 3 of Section 3.1. We will also use ∇ to denote these semi-conections.

For $w \in E_x$, set

$$Z^w(y) = X(y)Y_x(w).$$

Theorem 3.3.2. *Assume the diffusion operator $\mathcal{A}$ given by (3.9) is cohesive and let $\mathcal{B}$ be the operator on GLM determined by $\mathcal{A}$. Let E be the image of $\sigma^{\mathcal{A}}$, a vector bundle.*

(a) *The semi-connection ∇ induced by $\mathcal{B}$ is the adjoint of $\breve{\nabla}$ given by (3.10). Consequently $\nabla_w V = L_{Z^w} V$ for any vector field V and $w \in E_x$,*

(b) *For $u \in GLM$, identifying $\mathfrak{gl}(n)$ with $\mathcal{L}(\mathbf{R}^n; \mathbf{R}^n)$,*

$$\alpha(u) = \frac{1}{2} \sum \left(u^{-1}(-)\breve{\nabla}_{u(-)} X^p \right) \otimes \left(u^{-1}(-)\breve{\nabla}_{u(-)} X^p \right),$$

$$\beta(u) = -\frac{1}{2} \sum u^{-1} \breve{\nabla}_{\breve{\nabla}_{u(-)} X^p} X^p - \frac{1}{2} u^{-1} \mathrm{Ric}^{\#} u(-) + u^{-1} \breve{\nabla}_{u(-)} A.$$

Here $\check{\mathrm{Ric}}^{\#} : TM \to E$ is the Ricci curvature of $\check{\nabla}$ considered as an operator from TM to E, defined by

$$\check{\mathrm{Ric}}^{\#}(v) = \sum_{j=1}^{m} \check{R}(v, X^j(x))X^j(x)$$

for $\check{R}$ the curvature operator of $\check{\nabla}$.

Proof. The first part can be deduced from the stochastic flow results in Chapter 8 but we give a direct proof here. Let π_t^e be the flow of $X(\cdot)(e)$. It induces a linear map $\tilde{X}(u) : \mathbf{R}^m \to T_u GLM$ on the general linear bundle GLM:

$$\tilde{X}(\cdot)e = [X(\cdot)(e)]^{GL},$$

$$\tilde{X}(u)(e) = \frac{d}{dt}(TS_t^e \circ u)|_{t=0}, \qquad u \in GLM.$$

We can apply Lemma 2.3.1 with $\tilde{\mathbf{R}}^m = \mathbf{R}^m$ and so $\ell_u = Y(p(u))X(p(u))$. If $x = p(u)$ and $e \perp \ker[X(x)]$, then the horizontal lift map h_u defined by Theorem 3.1.2 is

$$h_u\left(X(x)(e)\right) = \tilde{X}(u)\left(\ell_u(e)\right) = \frac{d}{dt}\bigg|_{t=0}(T\pi_t^e \circ u). \qquad (3.11)$$

Note this will not hold in general if $e \in \ker[X(x)]$.

Let $\sigma : [0, T] \to M$ be a C^1 curve with $\dot{\sigma}(t) \in E_{\sigma(t)}$ each t. Then

$$Z^{\dot{\sigma}(t)}(x) := X(x)Y_{\sigma(t)}\dot{\sigma}(t).$$

Let $S_{s,t}^{\sigma}$ be the flow, from time s to time t, of the time dependent vector field $Z^{\dot{\sigma}(t)}$. Now $S_{s,t}^{\sigma}(\sigma(s)) = \sigma(t)$ for $0 \leqslant s \leqslant t \leqslant T$. Also, for any torsion free connection and any $v \in T_{\sigma(s)}M$,

$$\frac{D}{dt}\bigg|_{t=s} TS_{s,t}^{\sigma}(v) = \nabla Z^{\dot{\sigma}(t)}\left(TS_{s,t}^{\sigma}(v)\right)|_{t=s} = \nabla_v Z^{\dot{\sigma}(s)}.$$

Thus

$$\frac{D}{dt}TS_{0,t}^{\sigma}(v) = \nabla_{TS_{0,t}^{\sigma}} Z^{\dot{\sigma}(t)}.$$

If ϖ is the connection form of this torsion free connection, then

$$\varpi\left(\frac{D}{dt}TS_{0,t}^{\sigma} \circ u_0\right) = [e \mapsto (TS_{0,t}^{\sigma} \circ u_0)^{-1}\frac{D}{dt}TS_{0,t}^{\sigma}(u_0(e))]$$

$$= [e \mapsto (TS_{0,t}^{\sigma} \circ u_0)^{-1}\nabla_{TS_{0,t}^{\sigma}u_0(e)}Z^{\dot{\sigma}(t)}]$$

$$= \varpi\left(h_{TS_{0,t}^{\sigma} \circ u_0}(\dot{\sigma}(t))\right)$$

by (3.11), showing that the vertical parts of $\frac{d}{dt}\left(TS^\sigma_{0,t}\circ u_0\right)$ and $h_{TS^\sigma_{0,t}\circ u_0}(\dot\sigma(t))$ equal.

On the other hand, using this auxiliary connection, the horizontal parts of $\frac{d}{dt}\left(TS^\sigma_{0,t}\circ u_0\right)$ and $h_{TS^\sigma_{0,t}\circ u_0}(\dot\sigma(t))$ are both equal to the horizontal lift of $\dot\sigma(t)$. Thus

$$\frac{d}{dt}\left(TS^\sigma_{0,t}\circ u_0\right) = h_{TS^\sigma_{0,t}\circ u_0}(\dot\sigma(t))$$

and so $\{TS^\sigma_{0,t}\circ u_0 : 0 \leqslant t \leqslant T\}$ is the horizontal lift of $\{\sigma(t) : 0 \leqslant t \leqslant T\}$ with respect to the semi-connection induced by $\mathcal{B}$. However by Lemma 1.3.4 in Elworthy-LeJan-Li [36], $TS^\sigma_{0,t}(v)$ of $S^\sigma_{0,t}$ is the parallel translation of v along σ by the adjoint semi-connection $\hat{\nabla}$ of the LeJan-Watanabe connection on E associated to X and $\{TS^\sigma_{0,t}\circ u_0 : 0 \leqslant t \leqslant T\}$ is the horizontal lift of $\{\sigma(t) : 0 \leqslant t \leqslant T\}$ with respect to $\hat{\nabla}$. This proves the first claim. And $\nabla_w V = L_{Z^w}V$ by Lemma 1.3.4 of Elworthy-LeJan-Li [36].

For the last part let $\varpi : H \oplus VTGLM \to \mathfrak{g} = L(\mathbf{R}^n;\mathbf{R}^n)$ be the semi-connection 1-form. For $u_0 \in GLM$, set $u_t = T\xi_t \circ u_0$ where $\{\xi_t\}$ is a local flow for the stochastic differential equation

$$dx_t = X(x_t) \circ dB_t + A(x_t)dt \tag{3.12}$$

on M where $\{B_t\}$ is a Brownian motion on $\mathbf{R}^m$. (This defines the *derivative flow* on GLM.)

As for ordinary connections

$$\varpi(\circ du_t) = u_t^{-1}\frac{\hat{D}}{dt}(u_t-) \in \mathcal{L}(\mathbf{R}^n;\mathbf{R}^n).$$

Here, on the right-hand side u_t is differentiated as a process of linear maps $u_t \in \mathcal{L}(\mathbf{R}^n;T_{x_t}M)$ over (x_t). [It suffices to check the equality for C^1 curves (u_t) with $x_t = \pi(u_t)$ having $\dot{x}_t \in E_{x_t}$, $t \geqslant 0$. For this we can write $u_t = \tilde{x}_t \cdot g_t$ for $\tilde{x}_t$ a horizontal lift of $\{x_t\}$ and $g_t \in G$. Then observe that $\frac{\hat{D}}{dt}(u_t-) = \tilde{x}_t\frac{d}{dt}(\tilde{x}_t^{-1}u_t-).$] However as in [36],

$$u_t^{-1}\frac{\hat{D}}{dt}(u_t-) = u_t^{-1}\check{\nabla}_{u_t-}X \circ dB_t + u_t^{-1}\check{\nabla}_{u_t-}Adt.$$

From this the formula for $\alpha(u)$ follows by Remark 3.2.2(b). For $\beta(u)$ we need to identify the bounded variation part of $\int_0^t \varpi(\circ du_t)$. For this write

$$u_t^{-1}\check{\nabla}_{u_t-}X \circ dB_t = u_0^{-1}T_{x_0}\xi_t^{-1}\mathbin{/\!\!/_t} \circ \hat{\mathbin{/\!\!/}}_t^{-1}\check{\nabla}_{T\xi_t\circ u_0}X \circ dB_t$$

where $\hat{\mathbin{/\!\!/}}_t$ is the parallel translation along $\{\xi_s(x_0) : 0 \leqslant s \leqslant t\}$ using our semi-connection, which is the adjoint of $\check{\nabla}$ by Theorem 3.3.2. As in [36]

$$\hat{\mathbin{/\!\!/}}_t^{-1}\check{\nabla}_{T\xi_t\circ u_0}X \circ dB_t = \hat{\mathbin{/\!\!/}}_t^{-1}\check{\nabla}_{T\xi_t u_0}XdB_t - \frac{1}{2}\hat{\mathbin{/\!\!/}}_t^{-1}\mathrm{Ric}^\#(T\xi_t \circ u_0-)dt$$

while

$$u_0^{-1} T\xi_t^{-1} \widehat{/\!/}_t = u_0^{-1} - \int_0^t u_0^{-1} T\xi_s^{-1} \check{\nabla}_{/\!/_s-} X \circ dB_s - \int_0^t u_0^{-1} T\xi_s^{-1} \check{\nabla}_{/\!/_s-} A ds$$

giving the formula claimed for β. $\square$

Example: Gradient Brownian SDE

An isometric immersion $j : M \to \mathbf{R}^m$ of a Riemannian manifold M determines a stochastic differential equation on M:

$$dx_t = X(x_t) \circ dB_t$$

where $X(x) : \mathbf{R}^m \to T_x M$ is the orthogonal projection and $B.$ is a Brownian motion on $\mathbf{R}^m$. More precisely

$$X(x)(e) = \nabla[y \mapsto \langle j(y), \rangle](x).$$

It is well known that the solutions of the SDE are Brownian motions on M, see [30], [92], [31], and the equation is often called a "gradient Brownian SDE" . Moreover the LW connection given by equation (3.10) is the Levi-Civita connection, (by the classical construction of the latter), see [36]. Since the adjoint of the Levi-Civita connection is itself, Theorem 3.3.2, shows that our connection induced on GLM by the derivative flow of a gradient Brownian system is also the Levi-Civita connection. Almost by definition,

$$\langle \nabla_v X^p, w \rangle_{\mathbf{R}^m} = \langle \mathbf{a}(v, w), e_p \rangle_{\mathbf{R}^m} \tag{3.13}$$

where $\mathbf{a} : TM \times TM \to \mathbf{R}^m$ is the *second fundamental form* of the immersion with

$$\nabla_v X(e) = \mathbf{A}(v, n_x e) \qquad v \in T_x M, x \in M, e \in \mathbf{R}^m \tag{3.14}$$

for $n_x : \mathbf{R}^m \to T_x M^\perp$ the projection and $\mathbf{A} : TM \oplus TM^\perp \to TM$ the *shape operator* given by

$$\langle \mathbf{A}(v, e), w \rangle_{\mathbf{R}^m} = \langle \mathbf{a}(v, w), e_p \rangle_{\mathbf{R}^m}.$$

Here $TM^\perp$ refers to the normal bundle of M and $T_x M^\perp$ to the normal space at x to M, though we are considering its elements as being in the ambient space $\mathbf{R}^m$. Thus the vertical operator in the decomposition of the generator of the derivative flow on GLM for gradient flows is given by Theorem 3.3.2 with

$$\alpha(u) = \frac{1}{2} \sum_{j=1}^{m-n} u^{-1} \mathbf{A}(u-, l^j) \otimes u^{-1} \mathbf{A}(u-, l^j),$$

$$\beta(u) = -\frac{1}{2} \sum_{j=1}^{m-n} \mathbf{A}(\mathbf{A}(u-, l^j), l^j) - \frac{1}{2} u^{-1} Ric^{\#}(u-)$$

at a frame u over a point x. Here $l^1, \ldots, l^{m-n}$ denotes an orthonormal base for $T_x M^{\perp}$.

For the standard embedding of S^n in $\mathbf{R}^{n+1}$ we have

$$\mathbf{a}(u, v) = \langle u, v \rangle x$$

for $u, v \in T_x S^n$. Also the Ricci curvature is given by $Ric^{\#}(v) = (n-1)v$ for all $v \in TM$. Thus for the standard gradient SDE on S^n, at any frame u we have

$$\alpha(u) = \frac{1}{2} \mathrm{Id} \otimes \mathrm{Id} \tag{3.15}$$

$$\beta(u) = -\frac{1}{2} n \, \mathrm{Id}. \tag{3.16}$$

3.4   Associated Vector Bundles and Generalised Weitzenböck Formulae

As before let $\pi : P \to M$ be a smooth principal G-bundle and $\rho : G \to \mathbb{L}(V; V)$ a C^{∞} representation of G on some separable Banach space V. There is then the (possibly weakly) associated vector bundle $\pi^{\rho} : F \to M$ where $F = P \times V/\sim$ for the equivalence relation given by $(u, e) \sim (ug, \rho(g^{-1})e)$ for $u \in P$, $e \in V$, $g \in G$. If $[(u, e)] \in F$ denotes the equivalence class of (u, e) we can identify any $u \in P$ with a linear isomorphism

$$\bar{u} : V \to F_{\pi(u)}$$

by

$$\bar{u}(e) = [(u, e)]. \tag{3.17}$$

Consider the set of smooth maps from P to V which are equivariant by ρ:

$$M_{\rho}(P; V) = \{\text{smooth } Z : P \to V, Z(ug) = \rho(g)^{-1} Z(u), \ u \in P, g \in G\}.$$

There is the standard bijective correspondence $\mathfrak{F}^{\rho}$ between $M_{\rho}(P, V)$ and $\Gamma(F)$, the space of smooth sections of F defined by

$$\mathfrak{F}^{\rho}(Z)(x) = \bar{u}[Z(u)], \qquad u \in \pi^{-1}(x), Z \in M_{\rho}(P; V).$$

Via this map, an equivariant diffusion generator $\mathcal{B}$ on P induces a differential operator $\mathcal{B}^{\rho} \equiv \mathfrak{F}^{\rho}(\mathcal{B})$ on $\Gamma(F)$, of order at most 2, by

$$\mathfrak{F}^{\rho}(\mathcal{B})(\mathfrak{F}^{\rho}(Z)) = \mathfrak{F}^{\rho}[\mathcal{B}(Z)], \quad Z \in M_{\rho}(P; V). \tag{3.18}$$

Here $\mathcal{B}$ has been extended trivially to act on V-valued functions. Note that the definition makes sense since,

$$\mathcal{B}(Z)(ug) = \mathcal{B}\left(Z \circ R_g\right)(u) = \mathcal{B}\left(\rho(g)^{-1}Z\right)(u) = \rho(g)^{-1}\mathcal{B}(Z)(u).$$

For such a representation ρ let

$$\rho_* : \mathfrak{g} \to \mathcal{L}(V; V)$$

be the induced representation of the Lie algebra $\mathfrak{g}$ (the derivative of ρ at the identity).

Theorem 3.4.1. *When $\mathcal{B}$ is a vertical equivariant diffusion generator the induced operator on sections of any associated vector bundle is a zero-order operator. With the notation of Theorem 3.2.1, the zero-order operator in $\Gamma(F)$ induced by $\mathcal{B}$ is represented by $\lambda^\rho : P \to \mathcal{L}(V; V)$ for*

$$\lambda^\rho(u) = \rho_*(\beta(u)) + \mathrm{Comp}\left((\rho_* \otimes \rho_*)(\alpha(u))\right), \qquad u \in P \qquad (3.19)$$

for $\mathrm{Comp} : \mathcal{L}(V; V) \otimes \mathcal{L}(V; V) \to \mathcal{L}(V; V)$ *the composition map* $A \otimes B \mapsto AB$.

Proof. The operator $\mathcal{B}^\rho$ is a zero-order operator if $\mathcal{F}^\rho(\mathcal{B})(S)(x_0) = \mathcal{F}^\rho(\mathcal{B})(S')$ whenever two sections S and S' of F agree at x_0. This holds if $\mathcal{B}(fZ) = f\mathcal{B}(Z)$ for any invariant function $f : P \to R$ and V-valued function Z on P. But this holds by Remark 1.4.8.

For the representation (3.19), suppose $Z : P \to V$ is equivariant:

$$Z(u \circ g) = \rho(g)^{-1}Z(u), \qquad g \in G.$$

Then

$$\begin{aligned}
\mathcal{L}_{A_j^*}(Z)(u) &= \frac{d}{dt}\, Z(u \cdot e^{A_j t}))|_{t=0} \\
&= \frac{d}{dt}\, \rho(e^{-A_j t})Z(u)|_{t=0} \\
&= -\rho_*(A_j)Z(u).
\end{aligned}$$

Iterating we have

$$\mathcal{B}(Z)(u) = \sum \alpha^{ij}(u)\rho_*(A_j)\rho_*(A_i)Z(u) + \sum \beta_k \rho_*(A_k)Z(u)$$

proving (3.19). $\qquad\qquad\square$

From this theorem we easily have the following estimate, which combined with the discussion in Section 3.5 below, when applied to the associated bundle $\wedge F$ to the orthonormal bundle, shows that the Weitzenböck curvature is positive if the curvature is.

Corollary 3.4.2. *If ρ is an orthogonal representation, i.e. $(\rho_*(\alpha))^* = -\rho_*(\alpha)$ for all $\alpha \in \mathfrak{g}$, then $\lambda^\rho(v, v) \leqslant 0$ for all $v \in V$.*

Proof. Write $\alpha = \sum_k \mu_k A_k \otimes A_k$ where $\{A_k\}$ is as in Remark 3.2.2(c). Then for $v \in F$,

$$\langle \text{Comp} \circ (\rho_* \otimes \rho_*)(\alpha(u))(v), v \rangle = \langle \sum \mu_k [\rho_* A_k]^2 (v), v \rangle$$
$$= -\sum \mu_k \langle \rho_*(A_k)(v), \rho_*(A_k)(v) \rangle \leqslant 0,$$

since $\mu_k \geqslant 0$. The result follows from (3.19) since $\rho_*(\beta(u))$ is skew symmetric. $\quad\square$

The situation of Corollary 3.4.2 arises when considering the derivative flow for an SDE on a Riemannian manifold whose flow consists of isometries ; for example canonical SDE's on symmetric spaces as in Section 7.2 below and [36].

Example 3.4.3. We use the notation in Example 3.2.3. Let P be the special orthonormal frame bundle for E over the Heisenberg group H, with group S^1. We use the left action of the Heisenberg group to trivialise it to $H \times S^1$. Denote by $\rho : S^1 \to \mathcal{L}(\mathbf{R}^2; \mathbf{R}^2)$ the representation of S^1 given by $\rho(e^{it}) = \begin{pmatrix} \cos t & -\sin t \\ \sin t & \cos t \end{pmatrix}$. Let $s = s_1 X + s_2 Y$ be a section of $\Gamma(E)$, identified as $s_1 + is_2$. Let $\rho^*(s)$ be the induced equivariant map from $P \to R^2$:

$$\rho^*(s)(p, e^{i\theta}) = e^{-i\theta} s(p), \qquad\qquad p \in H.$$

Define

$$(\mathcal{B}^\rho s)(p) = e^{i\theta} \mathcal{B}(\rho^*(s))(p, e^{i\theta}).$$

As in Example 3.2.3 take

$$\mathcal{B} = \frac{1}{2}\left\{ (X^2 + Y^2) + r_1 \frac{\partial^2}{\partial x \partial \theta} + r_2 \frac{\partial^2}{\partial y \partial \theta} + \gamma \frac{\partial^2}{\partial^2 \theta} + (xr_2 - r_1 y)\frac{\partial^2}{\partial z \partial \theta} + (\frac{\partial r_1}{\partial x} + \frac{\partial r_2}{\partial y})\frac{\partial}{\partial \theta} \right\},$$

and recall that then

$$\mathcal{A}^H = \frac{1}{2}(\tilde{X}^2 + \tilde{Y}^2) \quad\text{and}\quad \mathcal{B}^V = \frac{1}{2}\left((\gamma - r_1^2 - r_2^2)\frac{\partial^2}{\partial^2 \theta} - \frac{1}{2}(x\frac{\partial r_2}{\partial z} - y\frac{\partial r_1}{\partial z})\frac{\partial}{\partial \theta} \right),$$

with semi-connection determined, as in Example 3.2.3, by the E-valued vector field V_0 given by

$$V_0(x, y, z) = -r_1 X - r_2 Y.$$

From this $\mathcal{B}^\rho = (\mathcal{B}^V)^\rho + (\mathcal{A}^H)^\rho$ with

$$\mathcal{B}^V(\rho^*(s))(p, e^{i\theta}) = -\frac{1}{2}(\gamma - r_1^2 - r_2^2)\rho^*(s)(p, e^{i\theta}) + i\frac{1}{4}(x\frac{\partial r_2}{\partial z} - y\frac{\partial r_1}{\partial z})\rho^*(s)(p, e^{i\theta})$$

and so

$$(\mathcal{B}^V)^\rho(s)(p) = -\frac{1}{2}(\gamma - r_1^2 - r_2^2)s(p) + i\frac{1}{4}(x\frac{\partial r_2}{\partial z} - y\frac{\partial r_1}{\partial z})s(p). \qquad (3.20)$$

We leave as an exercise the computation which verifies that

$$(\mathcal{A}^H)^p(s) = \frac{1}{2}\operatorname{trace}\hat{\nabla}_-\hat{\nabla}_-(s) \tag{3.21}$$

$$= \frac{1}{2}\left(\triangle^L s + i\operatorname{trace}\langle\nabla^L_- V_0, -\rangle_E + 2i\nabla^L_{V_0}s - |V_0|^2 s\right), \tag{3.22}$$

recalling from Example 3.2.3 that the covariant derivative of our connection is given by

$$\hat{\nabla}_w U = \nabla^L_w U + i\langle V_0(x, y, z), w\rangle_E U(x, y, z), \qquad w \in E_{(x,y,z)}$$

in terms of the left-invariant covariant derivative. Also $\triangle^L$ denotes trace $\nabla^L_-\nabla^L_-$.
    Let us relate this to Theorem 3.4.1. Let

$$\rho_* : t \in \mathfrak{s}^1 \to \begin{pmatrix} 0 & 1 \\ -1 & 0 \end{pmatrix} t$$

be the induced representation on the Lie algebra of S^1. Consider $\mathcal{B}^V$, the vertical part of $\mathcal{B}$. Now in the notation of Theorem 3.2.1 $\alpha^{1,1}(u) = \frac{1}{2}(\gamma - r_1^2 - r_2^2)$, and $\beta(u) = \frac{1}{4}(x\frac{\partial r_2}{\partial z} - y\frac{\partial r_1}{\partial z})$ and hence

$$\lambda^p(u) = \frac{1}{2}(\gamma - r_1^2 - r_2^2)\begin{pmatrix} 0 & 1 \\ -1 & 0 \end{pmatrix}\begin{pmatrix} 0 & 1 \\ -1 & 0 \end{pmatrix} + \frac{1}{4}(x\frac{\partial r_2}{\partial z} - y\frac{\partial r_1}{\partial z})\begin{pmatrix} 0 & 1 \\ -1 & 0 \end{pmatrix}$$

$$= -\frac{1}{2}(\gamma - r_1^2 - r_2^2) + i\frac{1}{4}(x\frac{\partial r_2}{\partial z} - y\frac{\partial r_1}{\partial z})$$

which is indeed the multiplication operator appearing in formula (3.20) for $(\mathcal{B}^V)^p$.
    $\square$

We use the following conventions, as in [36]. Let V be an N-dimensional real inner product space. For $1 \leqslant i \leqslant n$,

$$a_1 \wedge \cdots \wedge a_n = \frac{1}{n!}\sum_\pi \operatorname{sgn}(\pi)a_{\pi(1)} \otimes \cdots \otimes a_{\pi(n)},$$

$$\iota_v(u_1 \wedge \cdots \wedge u_q) = \sum_{j=1}^q (-1)^{j+1}\langle v, u_j\rangle u_1 \wedge \cdots \wedge \widehat{u_j} \wedge \cdots \wedge u_q \tag{3.23}$$

$\langle \otimes a_i, \otimes b_i\rangle = n!\Pi_i\langle a_i, b_i\rangle$, and $\langle \wedge a_i, \wedge b_i\rangle = \det(\langle a_i, b_j\rangle)$. Let $\wedge V$ stand for the exterior algebra of V and a_j^* the "creation operator" on $\wedge V$ given by $a_j^* v = e_j \wedge v$ for $(e_1, \ldots, e_N)$ an orthonormal basis for $\wedge V$. Let a_j be its adjoint, the "annihilation operator" given by $a_j = \iota_{e_j}$. Note the commutation law:

$$a_i a_j^* + a_j^* a_i = \delta_{ij}. \tag{3.24}$$

If $A : V \to V$ is a linear map on V, there are the operators $\wedge A$ and $(d\Lambda)(A)$ on $\wedge V$, which restricted to $\wedge^p V$ are:

$$(d\Lambda^p)(A)(u_1 \wedge \cdots \wedge u_p) = \sum_1^p u_1 \wedge \cdots \wedge u_{j-1} \wedge Au_j \wedge u_{j+1} \wedge \cdots \wedge u_p,$$

and also

$$(\wedge^p A)(u_1 \wedge \cdots \wedge u_p) = Au_1 \wedge \cdots \wedge Au_p.$$

A useful formula for $A \in \mathcal{L}(V; V)$ is

$$d\Lambda(A) = \sum_{i,j} A_{ij} a_i^* a_j. \tag{3.25}$$

Note that since $\alpha(u)$ is symmetric, $(\rho_* \otimes \rho_*)\alpha(u) : V \otimes V \to V \otimes V$ has

$$(\rho_* \otimes \rho_*)\alpha(u)(v^1 \wedge v^2) = \sum_{i,j} \alpha^{ij}(u)\rho_*(A_i) \otimes \rho_*(A_j)(v^1 \wedge v^2) \tag{3.26}$$

$$= \sum_{ij} \alpha^{ij}(u)\, \rho^*(A_i)v^1 \wedge \rho^*(A_j)v^2. \tag{3.27}$$

and so $(\rho_* \otimes \rho_*)\alpha(u)$ restricts to a map of $\wedge^2 V$ to itself.

Quantitative estimates can be obtained by some representation theory. For example suppose $G = O(n)$ with ρ the standard representation on $\mathbf{R}^n$. Consider the representation $\wedge^k \rho$ on $\wedge^k \mathbf{R}^n$.

Corollary 3.4.4. *Take the Hilbert-Schmidt inner product on $\mathfrak{so}(n)$ and let $0 \leqslant \mu_1(x) \leqslant \cdots \leqslant \mu(x)_{\frac{1}{2}n(n-1)}$ be the eigenvalues of α on the fibre $p^{-1}(x)$, $x \in M$, as described in Remark 3.2.2(c). Then for all $V \in \wedge^k \mathbf{R}^n$,*

$$-\frac{1}{2}k(n-k)\mu_{\frac{1}{2}n(n-1)}(x) \leqslant \left\langle \lambda^{\wedge^k}(u)V, V \right\rangle \leqslant -\frac{1}{2}k(n-k)\mu_1(x).$$

Proof. Following Humphreys [52], §6.2, consider the bilinear form β on $\mathfrak{so}(n)$ given by

$$\beta(A, B) = \mathrm{trace}\left((d\wedge^k)(A)(d\wedge^k)(B)\right) = \frac{(n-2)!}{(k-1)!(n-k-1)!}\, \mathrm{trace}(AB)$$

by a short calculation using elementary matrices. By Remark 3.2.2(c) since our inner product on $\mathfrak{so}(n)$ is $\mathrm{ad}(O(n))$-invariant we can write

$$\alpha(u) = \sum_{l=1}^{\frac{1}{2}n(n-1)} \mu_l(x)A_l(u) \otimes A_l(u)$$

with $x = p(u)$ and $\{A_l(u)\}_l$ an orthonormal base for $\mathfrak{so}(n)$ at each $u \in P$.

For each $u \in P$, set

$$A'_l(u) = \frac{(k-1)!(n-k-1)!}{(n-2)!} A_l(u)$$

to ensure $\beta(A'_l(u), A_j(u)) = \delta_{lj}$ for each u.
Then

$$\left\langle \mathrm{Comp} \circ (\rho_*^{\wedge^k} \otimes \rho_*^{\wedge^k})(\alpha(u))V, V \right\rangle = \sum \mu_l(x)\left\langle (d\wedge^k)A_l(u) \circ (d\wedge^k)A_l(u)V, V \right\rangle$$

$$= \sum_l \left[\frac{(k-1)!(n-k-1)!}{(n-2)!} \right]^{-1} \left\langle (d\wedge^k)A_l(u) \circ (d\wedge^k)A'_l(u)V, V \right\rangle$$

$$\leqslant - \frac{(n-2)!}{(k-1)!(n-k-1)!}\langle c_{\wedge^k}V, V\rangle,$$

where

$$c_{\wedge^k} = (d\wedge^k)A_l(u) \circ (d\wedge^k)A'_l(u),$$

the Casimir element of our representation $d\wedge^k$ of $\mathfrak{so}(n)$. Since the representation is irreducible, (for example see [12] Theorem 15.1 page 278), this element is a scalar, and we have, see Humphreys [52],

$$c_{\wedge^k} = \frac{\dim \mathfrak{so}(n)}{\dim \wedge^k \mathbf{R}^n} = \frac{1}{2}n(n-1)\Big/\frac{n(n-1)\dots(n-k+1)}{k!}.$$

Thus $\lambda^{\wedge^k}(u) \leqslant -\frac{1}{2}k(n-k)\mu_1$. The lower bound follows in the same way. $\qquad\square$

When $\mathcal{B}$ has an equivariant Hörmander form representation the zero-order operator $\mathcal{F}^\rho(V)$ can be given in a simple way by (3.28) below. This was noted for the classical Weitzenböck curvature terms using derivative flows in Elworthy [32].

Proposition 3.4.5. *Suppose $\mathcal{B}$ lies over a cohesive operator $\mathcal{A}$ and has a smooth Hörmander form: $\mathcal{B} = \frac{1}{2}\sum \mathcal{L}_{Y^j}\mathcal{L}_{Y^j} + \sum \beta_k \mathcal{L}_{Y^0}$ with the vector fields Y^j, $j = 1,\dots,m$, being G-invariant. Let (η_t^j) be the flow of Y^j. For a representation ρ of G with associated vector bundle $\pi^\rho : F \to M$ the zero-order operator $\mathcal{F}^\rho(\mathcal{B}^V)$ corresponding to the vertical component of $\mathcal{B}$ is given by*

$$\mathcal{F}^\rho(\mathcal{B}^V)(x_0) = \frac{1}{2}\sum_{j=1}^m \frac{D^2}{dt^2}\eta_t^j(u_0)\Big|_{t=0} \circ (\bar{u}_0)^{-1} + \frac{D}{dt}\eta_t^0(u_0)\Big|_{t=0} \circ (\bar{u}_0)^{-1} \quad (3.28)$$

for any $u_0 \in \pi^{-1}(x_0)$.

Proof. Set $u_t^j = \eta_t^j(u_0) \in P$ and $\sigma(t) = \pi(u_t^j)$ so $\bar{u}_t^j \in \mathcal{L}(V; F_{\sigma(t)})$. From Remark 3.2.2(b),

$$\alpha(u_0) = \frac{1}{2}\sum_{j=1}^m \varpi(Y^j(u_0)) \otimes \varpi(Y^j(u_0))$$

and so

$$(\rho_* \otimes \rho_*)\alpha(u_0) = \frac{1}{2}\sum_{j=1}^{m}(\bar{u}_0)^{-1}\frac{D}{dt}\bar{u}_t^j\Big|_{t=0} \otimes (\bar{u}_0)^{-1}\frac{D}{dt}\bar{u}_t^j\Big|_{t=0}$$

as in the proof of Theorem 3.3.2.

Also from equation (3.6)

$$\beta(u_0) = \frac{1}{2}\sum_{j=1}^{m}\mathcal{L}_{Y^j}\big(\varpi(Y^j(-))\big)(u_0) + \frac{1}{2}\big(\varpi(Y^0(-))\big)(u_0).$$

Let $(/\!/_t)$ denote parallel translation in F along σ. Then

$$\begin{aligned}
\rho_*\mathcal{L}_{Y^j}\big(\varpi(Y^j(-))\big)(u_0) &= \frac{d}{dt}\rho_*\varpi\Big(Y^j(u_t^j)\Big)\Big|_{t=0} \\
&= \frac{d}{dt}\Big((\bar{u}_t^j)^{-1}\frac{D}{dt}\bar{u}_t^j\Big)\Big|_{t=0} \\
&= \frac{d}{dt}\Big((/\!/_t^{-1}\overline{u_t^j})^{-1}/\!/_t^{-1}\frac{D}{dt}\overline{u_t^j}\Big)\Big|_{t=0} \\
&= -\bar{u}_0^{-1}\frac{D}{dt}\overline{u_t^j}\Big|_{t=0} \circ \bar{u}_0^{-1}\frac{D}{dt}\overline{u_t^j}\Big|_{t=0} + \bar{u}_0^{-1}\frac{D^2}{dt^2}\overline{u_t^j}\Big|_{t=0}
\end{aligned}$$

leading to the required result via Theorem 3.4.1. $\square$

To examine particular examples we will need to have detailed information about the zero-order operators determined by a vertical diffusion generator. For this suppose $\mathcal{B}$ is vertical and given by

$$\mathcal{B} = \sum \alpha^{ij}\mathcal{L}_{A_i^*}\mathcal{L}_{A_j^*} + \sum \beta_k \mathcal{L}_{A_k^*}$$

for $\alpha : P \to \mathfrak{g} \otimes \mathfrak{g}$ and $\beta : P \to \mathfrak{g}$ as in Theorem 3.2.1 and (3.4).

Motivated by the Weitzenböck formula for the Hodge-Kodaira Laplacian on differential forms, see Corollary 3.4.9 below, [93], [22], we shall examine in more detail the case of the exterior power $\wedge\rho : G \to \mathbf{L}(\wedge V; \wedge V)$ of a fixed representation ρ showing that $\lambda^{\wedge\rho}$ has expressions in terms of annihilation and creation operators which are structurally the same as these of the Weitzenböck curvature (which are shown to be a special case in Corollary 3.4.9). For notational convenience we give V an inner product in what follows.

Lemma 3.4.6. *If $\mathcal{B}$ is a vertical operator on P and $(e_i, \ i = 1, 2, \ldots, N)$ is an orthonormal basis of V, the zero-order operator on the associated bundle $\wedge F \to M$ is represented by $\lambda^{\wedge\rho} : P \to \mathbf{L}(\wedge^p V; \wedge^p V)$ with*

$$\lambda^{\wedge\rho}(u) = \sum_{i,j,k,l=1}^{N} \langle ((\rho_* \otimes \rho_*)\alpha(u))(e_j \otimes e_l), e_i \otimes e_k \rangle\, a_i^* a_j a_k^* a_l$$

$$+ \sum_{i,j=1}^{N} \langle (\rho_*\beta(u))e_j, e_i \rangle a_i^* a_j, \qquad u \in P.$$

Proof. Recall that if $A \in \mathbf{L}(V; V)$, then

$$d\Lambda(A) = \sum_{i,j=1}^{N} \langle Ae_j, e_i \rangle a_i^* a_j, \tag{3.29}$$

e.g. see Cycon-Froese-Kirsch-Simon [22]. Consequently

$$d\Lambda(\rho_* \beta(u)) = \sum_{i,j=1}^{N} \langle \rho_* \beta(u) e_j, e_i \rangle a_i^* a_j. \tag{3.30}$$

On the other hand by Theorem 3.2.1 and (3.4), we can represent α as:

$$\alpha(u) = \sum_{n,m} a_{n,m}(u) A_n \otimes A_m$$

where $\{A_i\}_{i=1}^{N}$ is a basis of $\mathfrak{g}$. So

$$\text{Comp} \circ (\wedge \rho_* \otimes \wedge \rho_*)(\alpha(u))$$

$$= \text{Comp} \circ \sum_{m,n} a_{n,m}(u) \, d\Lambda(\rho_* A_m) \otimes d\Lambda(\rho_* A_n)$$

$$= \sum_{m,n} a_{n,m}(u) d\Lambda(\rho_* A_m) \circ d\Lambda(\rho_* A_n)$$

$$= \sum_{m,n} a_{n,m}(u) \sum_{i,j,k,l=1}^{N} \langle \rho_* A_m e_j, e_i \rangle \langle \rho_* A_n e_l, e_k \rangle a_i^* a_j a_k^* a_l$$

$$= \frac{1}{2} \sum_{m,n} a_{n,m}(u) \sum_{i,j,k,l=1}^{N} \langle (\rho_* A_m \otimes \rho_* A_n)(e_j \otimes e_l), e_i \otimes e_k \rangle \, a_i^* a_j a_k^* a_l$$

$$= \frac{1}{2} \sum_{i,j,k,l=1}^{N} \langle (\rho_* \otimes \rho_*) \alpha(u)(e_j \otimes e_l), e_i \otimes e_k \rangle \, a_i^* a_j a_k^* a_l,$$

since our convention for the inner product on tensor products gives

$$\langle u_1 \otimes v_1, u_2 \otimes v_2 \rangle = 2 \langle u_1, u_2 \rangle \langle v_1, v_2 \rangle.$$

The desired conclusion follows. □

Theorem 3.4.7. *Let* $R(u) : \wedge^2 V \to \wedge^2 V$ *be the restriction of* $2(\rho_* \otimes \rho_*)\alpha(u) : V \otimes V \to V \otimes V$, *then*

$$\lambda^{\wedge \rho}(u) = - \sum_{i < k, j < l} \langle R(u)(e_j \wedge e_l), e_i \wedge e_k \rangle \, a_i^* a_k^* a_j a_l$$

$$+ \sum_{i,j,l=1}^{N} \langle (\rho_* \otimes \rho_*) \alpha(u)(e_j \otimes e_l), e_i \otimes e_j \rangle \, a_i^* a_l + \sum_{i,j} \langle \rho_* \beta(u) e_j, e_i \rangle \, (a_i)^* a_j.$$

This can be rewritten as:

$$\lambda^{\wedge\rho}(u) = - \sum_{i<k,j<l} \langle R(u)(e_j \wedge e_l), e_i \wedge e_k \rangle \, a_i^* a_k^* a_j a_l + d \wedge (Z^\rho(u)) + d \wedge (\rho_* \beta(u)).$$

$$(3.31)$$

where $Z^\rho(u) \in \mathbf{L}(V; V)$ is defined by

$$\langle Z^\rho(v_1), v_2 \rangle = \sum_{j=1}^N \langle (\rho_* \otimes \rho_*)(\alpha(u))(e_j \otimes v_1), v_2 \otimes e_j \rangle_{V \otimes V}.$$

Proof. This follows from Lemma 3.4.6 using the anti-commutation formula (3.24) since

$$\sum_{i,j,k,l=1}^N \langle (\rho_* \otimes \rho_*)\alpha(u)(e_j \otimes e_l), e_i \otimes e_k \rangle \, a_i^* a_j a_k^* a_l$$

$$= - \sum_{i,j,k,l=1}^N \langle (\rho_* \otimes \rho_*)\alpha(u)(e_j \otimes e_l), e_i \otimes e_k \rangle \, a_i^* a_k^* a_j a_l$$

$$+ \sum_{i,j,l=1}^N \langle (\rho_* \otimes \rho_*)\alpha(u)(e_j \otimes e_l), e_i \otimes e_j \rangle \, a_i^* a_l$$

$$= - \sum_{j<l;i<k}^N \langle R(u)(e_j \wedge e_l), e_i \wedge e_k \rangle \, a_i^* a_k^* a_j a_l$$

$$+ \sum_{i,j,l=1}^N \langle (\rho_* \otimes \rho_*)\alpha(u)(e_j \otimes e_l), e_i \otimes e_j \rangle \, a_i^* a_l. \qquad \square$$

Remark 3.4.8. (a) Note that the second term in (3.31) in general depends on the symmetric part of $(\rho_* \otimes \rho_*)(\alpha(u))$ as well as on R.

(b) If we write

$$\alpha(u) = \sum \mu_k(u) A_k(u) \otimes A_k(u)$$

as in Remark 3.2.2(c), Then $Z^\rho(u)$ in (3.31) has

$$Z^\rho(u) = \sum_k \mu_k(u) \rho_*(A_k(u)) \rho_*(A_k(u)).$$

Corollary 3.4.9. *For the derivative process in GLM of a cohesive generator $\mathcal{A}$ given in Hörmander form without a drift, the zero-order operator induced by the vertical diffusion on the exterior bundles $\wedge TM$ is minus one half times the generalised Weitzenböck curvature, $\check{R}^* : \wedge^* TM \to \wedge^* E$, given by:*

$$\check{R}^q V = d \wedge^q (\mathrm{Ric}^\#)(V) - 2 \sum_{\substack{1 \leq i < k \leq n \\ 1 \leq j < l \leq p}} R_{ikjl} a_l^* a_j^* a_k a_i V \qquad (3.32)$$

for all $V \in \wedge^q TM$. Here $R_{ikjl} = \langle R(e_i, e_k)e_l, e_j \rangle, 1 \leqslant i, k \leqslant n, 1 \leqslant j, l \leqslant p$ for $R : TM \oplus TM \to \mathbb{L}(E, E)$ the curvature transform of the associated connection on E, and if $V \in \wedge^q T_x M$ the set $\{e_1, \ldots, e_p\}$ is an orthonormal base for E_x which together with $e_{p+1}, \ldots, e_n$ forms an orthonormal base for some inner product on $T_x M$ extending that of E_x.

Proof. By Theorem 3.3.2,

$$\alpha(u) = \frac{1}{2} \sum \left(u^{-1} \nabla_{u(-)} X^p\right) \otimes \left(u^{-1} \nabla_{u(-)} X^p\right), \qquad u \in GLM.$$

By Corollary C.5 in [36] the restriction $\frac{1}{2} R(u)$ of $\alpha(u)$ to anti-symmetric tensors corresponds to one half of the curvature operator $\frac{1}{2} \mathcal{R} : \wedge^2 TM \to \wedge^2 E$ composed with the inclusion of $\wedge^2 E$ into $\wedge^2 TM$.

By the relation between the curvature transform and the curvature operator:

$$\langle \mathcal{R}(v^1 \wedge v^2), w^1 \wedge w^2 \rangle = \langle R(v^1, v^2)w^2, w^1 \rangle,$$

the first term in $\lambda^p(u)$ of Theorem 3.4.7 corresponds to:

$$- \sum_{\substack{1 \leqslant i < k \leqslant n \\ 1 \leqslant j < l \leqslant p}} \langle \mathcal{R}(e_i \wedge e_k), e_j \wedge e_l \rangle \, a_j^* a_l^* a_i a_k = - \sum_{\substack{1 \leqslant i < k \leqslant n \\ 1 \leqslant j < l \leqslant p}} R_{iklj} a_j^* a_l^* a_i a_k$$

$$= \sum_{\substack{1 \leqslant i < k \leqslant n \\ 1 \leqslant j < l \leqslant p}} R_{ikjl} a_l^* a_j^* a_k a_i$$

by the skew-symmetry of R_{iklj} in i, k and in $j.l$ and the anti-commutation of annihilation operators. By (ii) of Remark 3.4.8, the second term corresponds to

$$\frac{1}{2} d \wedge^q \left(\sum_{j=1}^{m} \nabla_{\nabla_{-X^j} X^j} \right).$$

The required result follows since

$$\beta(u) = -\frac{1}{2} \sum_{j=1}^{m} u^{-1} \left(\nabla_{\nabla_{u(-)} X^j} X^j \right) - \frac{1}{2} u^{-1} \left(\mathrm{Ric}^{\#} u(-) \right). \qquad \square$$

Corollary 3.4.9 reflects the results in [36], Theorem 2.4.2, concerning Weitzenböck formulae for Hörmander form operators on differential forms. In particular it gives another approach to the result that when $E = TM$ and $\check{\nabla}$ is the Levi-Civita connection, as holds for gradient stochastic differential equations, the generator induced on differential forms by the derivative process is a constant times the Hodge-Kodaira Laplacian for the induced Riemannian structure on M up to a first-order term, see [36]. This comes from identifying the adjoint of $\check{R}^q$ given by equation (3.32) when $\check{\nabla}$ is the Levi-Civita connection, with the standard Weitzenböck term

R^q, say, which appears in the Weitzenböck formula for the Hodge-Kodaira Laplacian on q-forms:

$$\triangle^q \phi \equiv -(dd^* + d^*d)\phi = \nabla^* \nabla \phi - R^q \phi. \tag{3.33}$$

and is the zero-order operator given by

$$R^q \phi \equiv R^{\bullet} \phi = \sum R_{ijkl}(a^i)^* a^j (a^k)^* a^l \phi$$

see [22], or [93] where the opposite sign is used. Here the annihilation and creation operators acting on a form ϕ are denoted by a^i and $(a^i)^*$ so that $(a^i \phi)(v) = \phi(a_i^* v) = \phi(e_i \wedge v)$ and $((a^i)^* \phi)(v) = \phi(a_i v) = \phi(\iota_{e_i} v) = (e_i^* \wedge \phi)(v)$.

One of the standard ways of writing R^q is with the decomposition

$$R^q \phi = \phi \circ d \wedge^q (\mathrm{Ric}^{\#}) + \tilde{R}_{(4)} \phi$$

for ϕ a q-form, where

$$\tilde{R}_{(4)} = -\sum R_{ijkl}(a^i)^*(a^k)^* a^j a^l.$$

This follows from the definition of R^q using the anti-commutation relation (3.24) and the formula (3.25). For example see [22] page 260.

In [36] we showed that the second term in the right-hand side of equation (3.32) corresponds to $\tilde{R}_{(4)}$. There was an error in the sign of this term in the statement of Theorem 2.4.2 of [36] and in the discussion of its relationship with $\tilde{R}_{(4)}$, and for completeness we repeat the argument.

Working with the Levi-Civita connection and using the anti-commutatitvity of the creation operators we have

$$\tilde{R}_{(4)} \equiv - \sum_{i,j,k,l=1}^{n} R_{ijkl}(a^i)^*(a^k)^* a^j a^l$$

$$= - \sum_{1 \leqslant i < k}^{n} \sum_{j,l=1}^{n} [R_{ijkl} - R_{kjil}](a^i)^*(a^k)^* a^j a^l$$

$$= - \sum_{i<k} \sum_{j<l} [R_{ijkl} - R_{kjil} - R_{ilkj} + R_{klij}](a^i)^*(a^k)^* a^j a^l.$$

However by Bianchi's identity:

$$R_{ijkl} + R_{iljk} + R_{iklj} = 0$$

we have

$$R_{ijkl} - R_{kjil} = R_{ikjl},$$

and interchanging j and l:

$$-R_{ilkj} + R_{klij} = -R_{iklj} = R_{ikjl}.$$

Thus

$$\check{R}_{(4)} = -2 \sum_{i<k} \sum_{j<l} R_{ikjl}(a^i)^*(a^k)^* a^j a^l.$$

Taking adjoints we obtain the second term in the expression for $\check{R}^q V$ in equation (3.32), as claimed.

Note that if $\mathcal{B}$ is the operator on GLM determined by the Hörmander form (3.9) of $\mathcal{A}$, then for a representation $\rho : GL(M) \to \mathcal{L}(V;V)$ with associated $\pi^\rho : GL(n) \to \mathcal{L}(V;V)$ the induced operator $\mathcal{F}^\rho(\mathcal{B})$ on sections of π^ρ is also given by the 'Hörmander form' $\frac{1}{2} \sum_j \mathcal{L}_{X^j} \mathcal{L}_{X^j} + \mathcal{L}_A$, where for any C^1 vector field Y on M and any C^1 section U of π^ρ the Lie derivative $\mathcal{L}_Y U \in \Gamma F$ is given by

$$(\mathcal{L}_Y U)(x) = \bar{\mathbf{u}} \frac{d}{dt} \left(\mathbf{T}\eta_t^{\mathbf{Y}} \circ \mathbf{u} \right)^{-1} U \left(\eta_t^Y(x) \right) \Big|_{t=0}$$

for $x \in M$, u a frame at x, and (η_t^Y) the flow of Y, using the notation of (3.17). Indeed by (3.17), for $Z(u) = \bar{u}U(\pi(u))$, so $U = \mathcal{F}^\rho(Z)$,

$$\mathcal{F}^\rho(\mathcal{B})(U) = \mathcal{F}^\rho \left[\left(\frac{1}{2} \sum_j \mathcal{L}_{(X^j)^{GL}} \mathcal{L}_{(X^j)^{GL}} + \mathcal{L}_{A^{GL}} \right)(Z) \right]$$

while $\mathcal{L}_{(X^j)^{GL}}(Z)(u) = \frac{d}{dt} Z(T\eta_t^{X^j} \circ u) \Big|_{t=0}$ so that

$$\mathcal{F}^\rho \left[\mathcal{L}_{(X^j)^{GL}}(Z) \right](x) = \bar{\mathbf{u}} \frac{d}{dt} Z(T\eta_t^{X^j} \circ u) \Big|_{t=0} = \mathcal{L}_{X^j}(U)(x).$$

This representation of $\mathcal{F}^\rho(\mathcal{B})$ was noted in the case of the operator induced on differential forms by a stochastic flow in [36], and for the case of the Hodge-Kodaira Laplacian in Elworthy [32], see also Kusuoka [60].

Remark 3.4.10 (Elementary matrices, $\wedge^2 \mathbf{R}^n$ and $\mathfrak{so}(n)$). We always give $\mathfrak{so}(n)$ its Hilbert-Schmidt inner product $\langle A, B \rangle = \text{trace } B^* A$. Using the standard basis $e_1, \ldots, e_n$ of R^n let $E_{[p,q]}$ denote the elementary matrix $E^n_{[p,q]}$ with $E_{[p,q]}(v) = v_q e_p$. As in Example 3.0.1 set $A_{[p,q]} = \frac{1}{\sqrt{2}}[E_{[p,q]} - E_{[q,p]}]$ so that $\{A_{[p,q]} : 1 \leqslant p < q \leqslant n\}$ forms an orthonormal basis for $\mathfrak{so}(n)$. We have the isomorphism, but not isometry, of $\wedge^2 \mathbf{R}^n$ with $\mathfrak{so}(n)$ given by mapping $e_p \wedge e_q$ to $\frac{1}{2}(E_{[p,q]} - E_{[q,p]})$ in accordance with our usual interpretation of elements of $\mathbf{R}^n \otimes \mathbf{R}^n$ as linear operators on R^n. Then $\sqrt{2}e_p \wedge e_q$ corresponds to $A_{[p,q]}$.

Example 3.4.11. Let P be the orthonormal frame bundle for a Riemannian metric on M. Let $C : P \to \mathbf{L}(\mathbf{R}^n; \mathbf{R}^n)$ satisfy $C(ug) = g^{-1}C(u)g$ for $g \in O(n)$ with $C(u)$ a symmetric map for all $u \in P$. Define $\alpha : P \to \mathfrak{so}(n) \otimes \mathfrak{so}(n)$ by

$$\alpha(u) = \sum_{\substack{1 \leqslant p < q \leqslant n \\ 1 \leqslant p' < q' \leqslant n}} \text{trace}\langle C(u)A_{[p,q]} -, A_{[p',q']} -\rangle_{\mathbf{R}^n} A_{[p,q]} \otimes A_{[p',q']},$$

where the $\{A_{[p,q]}\}$ are as in Remark 3.4.10 above. Then α corresponds to an equivariant operator B^V, say, on P, namely, at $u \in P$,

$$B^V = \sum_{\substack{1 \leqslant p < q \leqslant n \\ 1 \leqslant p' < q' \leqslant n}} \text{trace}\langle C(u)A_{[p,q]}-, A_{[p',q']}-\rangle_{\mathbf{R}^n} \mathcal{L}_{A^*_{[p,q]}} \mathcal{L}_{A^*_{[p',q']}},$$

and

$$\text{Comp} \circ \alpha(u) = -\frac{1}{4}(\text{trace } C(u))\text{id} + \frac{1}{4}(2-n)C(u).$$

Let $\text{Ric}^{\#} : TM \to TM$ be the Ricci curvature (for the Levi-Civita connection, say)and take $C(u) = u \, \text{Ric}^{\#}_{\pi(u)}(u^{-1}-)$ for $u \in P$. If the Ricci curvature is non-negative the operator B^V is a vertical diiffusion operator whose induced zero-order term on vector fields is $\frac{1}{4}(2-n) \, \text{Ric}^{\#}_{\pi(u)} - \frac{1}{4}k$, where k is the scalar curvature.

Proof. Since $\alpha(ug) = (ad(g) \otimes ad(g)) \, \alpha(u)$ for $g \in O(n)$ the operator B^V is equivariant. To compute $\text{Comp} \circ \alpha(u)$ first observe that

$$\text{trace}\langle C(u)A_{[p,q]}-, A_{[p',q']}-\rangle_{\mathbf{R}^n} = \frac{1}{2}\langle C(u)A_{[p,q]} + A_{[p,q]}C(u), A_{[p',q']}\rangle_{\mathfrak{so}(n)}$$

and so, because the $\{A_{[p,q]}\}$ form an orthonormal base,

$$\alpha = \frac{1}{2}\sum \{CA_{[p,q]} \otimes A_{[p,q]} + A_{[p,q]}C \otimes A_{[p,q]}\}.$$

Then we use the elementary fact about elementary matrices:

$$E_{[p,q]}CE_{[p',q']} = C_{qp'}E_{[p,q']} \qquad \qquad \square$$

3.5 Notes

G-invariant diffusion operators

Suppose a diffusion operator on N is invariant under the action of a Lie group G. Even if the action is not principal in the sense that the quotient mapping $N \to N/G$ is not a principal bundle it may be possible to classify the orbits and give partial skew-product decompositions. This is discussed in Lázaro-Cami & Ortega's article,[62], with special reference to stochastic Hamiltonian systems, and their analysis could easily be combined with the use of the connections described here.

Canonical vertical diffusions on GLM

We have seen in Corollary 3.4.9 that there is a zero-order operator on the associated bundle $\wedge F \to M$ represented by the Weitzenböck curvature of a given connection. On the other hand, given a curvature operator $\mathcal{R}$ of a metric connection, or more generally an operator which has the same symmetry properties as a curvature tensor, is there a canonical vertical diffusion operator on GLM which induces zero-order operators on differential forms which have the form of the Weitzenböck curvatures of $\mathcal{R}$? A vertical operator with such a zero-order term always exists since we can take $\mathcal{R}$ in a diagonal form:

$$\mathcal{R}(u) = \sum_{n=1}^{N} A_n(u) \wedge A_n(u), \qquad (3.34)$$

for some $A_n : GLM \to \mathfrak{gl}(n)$ which are ad(G)-invariant, e.g. by taking an isometric embedding (*e.g.* see [36]). In this case let (e^j) be a basis of $E_{\pi(u)}$ and define

$$\alpha(u) = \frac{1}{2} \sum_{n=1}^{N} A_n(u) \otimes A_n(u),$$

$$\beta(u) = -\frac{1}{2} \sum_{n=1}^{N} (A_n(u))^2 - \frac{1}{2} \sum_{j=1}^{p} R(-, e^j) e^j, \qquad (3.35)$$

see Remark 3.4.8(b). Then α is positive and we can define an operator with its coefficients these α and β.

For a discussion of the representation of $\mathcal{R}$ in the form of (3.34) see Kobayashi-Nomizu [56] (Notes 17 and 18). In particular there is a discussion there of the number N required and of a rigidity theorem originating from Chern. See also Berger-Bryant-Griffiths [10].

When M is Riemannian with positive semi-definite curvature operator $\mathcal{R} : \wedge^2 TM \to \wedge^2 TM$ there is a canonical construction. For this take the orthonormal frame bundle $\pi : OM \to M$, with $G = O(n)$. We will use the isomorphism of $\wedge^2 \mathbf{R}^n$ with $\mathfrak{so}(n)$ described in Remark 3.4.10. Define

$$\alpha : OM \to \mathfrak{so}(n) \otimes \mathfrak{so}(n)$$

by

$$\alpha(u) = \frac{1}{4} \sum_{1 \leqslant p \leqslant q \leqslant n, 1 \leqslant p' \leqslant q' \leqslant n} \Big\langle \mathcal{R}(\wedge^2(u)(e_p \wedge e_q)), \wedge^2(u)(e_{p'} \wedge e_{q'})) \Big\rangle_{\pi(u)} A_{[p,q]} \otimes A_{[p',q']}.$$

Our representation ρ is just the identity map and, by (3.27) and Bianchi's identity, the restriction of $\alpha(u) : \mathbf{R}^n \otimes \mathbf{R}^n \to \mathbf{R}^n \otimes \mathbf{R}^n$ to $\wedge^2 \mathbf{R}^n$ is just $\frac{1}{2}\mathcal{R}$. In the notation of (3.31) we see

$$\langle Z^\rho(v^1), v^2 \rangle = -\frac{1}{2} \operatorname{Ric}(v^1, v^2).$$

If we take $\beta = 0$, we obtain from (3.31) that

$$\lambda^{\wedge^\rho}(u) = -\sum_{l < k, j < l} R_{jlik} a_i^* a_k^* a_j a_l - 2 (d\wedge) \operatorname{Ric}^\#.$$

To get the full Weitzenböck term, extend α over GLM by equivariance and define $\beta(u)$, for $u \in GLM$, by $\beta(u) = \frac{3}{2} u^{-1} \operatorname{Ric}^\#(u-)$ as in (3.35).

Chapter 4

Projectible Diffusion Processes and Markovian Filtering

Let M^+ be the Alexandrov one-point compactification, $M \bigcup \Delta$, of a smooth manifold M. Consider the space $\mathcal{C}M^+$ of processes (y_t) on M^+ with explosion time ζ, such that $t \to y_t$ is continuous with $y_t = \Delta$ when $t \geqslant \zeta$. We shall use $\mathcal{C}_{y_0}M^+$ to denote those processes starting at a given point y_0 of N. Let $\mathcal{L}$ be a diffusion operator on M and let $\{\mathbf{P}_{y_0}, y_0 \in M^+\}$ be the family of $\mathcal{L}$-diffusion measures in the sense of [53], *i.e.* the solution to the martingale problem on $\mathcal{C}M^+$ so the canonical process $(y_t, 0 \leqslant t < \zeta)$ with the system of diffusion measures $\{\mathbf{P}^{\mathcal{L}}_{y_0}, y_0 \in N^+\}$ is a strong Markov process on M^+. Denote by $\mathbf{E}$ mathematical expectation with respect to the measure $\mathbf{P}_{y_0}$. We may add to these notations the relevant subscripts or superscripts indicating the diffusion operator or the Markov process concerned, *e.g.* $\{\mathbf{P}^{\mathcal{L}}_{y_0}\}, \zeta^{\mathcal{L}}, \mathbf{E}^{\mathcal{L}, y_0}$ or even $\mathbf{E}^{y_0}$.

For $y_0 \in M$ and $f \in \mathcal{C}_c^\infty M$, the space of smooth functions on M with compact support, let

$$M_t^{df} := M_t^{df, \mathcal{L}} := f(y_{t \wedge \zeta}) - f(y_0) - \int_0^{t \wedge \zeta} \mathcal{L}f(y_s)ds. \tag{4.1}$$

Then $(M_t^{df} : 0 \leqslant t < \infty)$ is a martingale on the probability space $(\mathcal{C}(M), \mathbf{P}^{\mathcal{L}}_{y_0})$ with respect to the $\{\mathcal{F}_t^{y_0}\}$, where $\mathcal{F}_t^{y_0} = \sigma\{y_s; 0 \leqslant s \leqslant t\}$. Moreover it has bracket

$$\langle M^{df} \rangle_t = 2 \int_0^{t \wedge \zeta} \sigma^{\mathcal{L}}((df)_{y_s}, (df)_{y_s})ds.$$

This definition extends to the case of C^2 functions f but then M_t^{df} is only defined for $0 \leqslant t < \zeta^{\mathcal{L}}$ and is a local martingale.

K.D. Elworthy et al., *The Geometry of Filtering*, Frontiers in Mathematics, DOI 10.1007/978-3-0346-0176-4_4, © Springer Basel AG 2010

4.1   Integration of predictable processes

Proposition 4.1.1. *Let τ be a stopping time with $\tau < \zeta$ and let $\{\alpha_t : 0 \leqslant t < \tau\}$ be an $\mathcal{F}^{y_0}_*$ predictable process in T^*M such that $\alpha_t \in T^*_{y_t}M$ for each $t \in [0, \tau)$, and for each compact subset C of M we have*

$$\int_0^\tau \chi_C(y_s)\alpha_s(\sigma^{\mathcal{L}}\alpha_s)\,ds < \infty$$

almost surely.

Then there is a unique local martingale $\{M^\alpha_t : 0 \leqslant t < \tau\}$ such that for all $f \in \mathcal{C}^\infty_c M$,

$$\langle M^\alpha, M^{df} \rangle_t = 2 \int_0^t \sigma^{\mathcal{L}}(\alpha_s, (df)_{y_s})\,ds, \qquad t < \zeta. \tag{4.2}$$

Proof. We can write

$$\alpha_t = \sum_{j=1}^m g^j_t \cdot df_j(y_t), \tag{4.3}$$

where the functions g^j are predictable real-valued processes, for example by taking $(f_1, \ldots, f_m) : M \to \mathbf{R}^m$ to be an embedding and $g^j_t = \alpha_t(X^j(y_t))$, for $X(x) = \sum_{i=1}^m X^i(x)e_i$ the projection from $\mathbf{R}^m$ to T_xM. Using a partition of unity, at the cost of having an infinite, but locally finite sum, we can assume that the f_j in the representation are all in $\mathcal{C}^\infty_c M$. Define

$$M^\alpha_t := \sum_j \int_0^t g^j_s dM^{df_j}_s. \tag{4.4}$$

Clearly (4.2) holds. For uniqueness suppose K is a local martingale orthogonal to M^{df} for all $f \in C^\infty_c M$.

Then K vanishes since the martingale problem for $\mathcal{L}$ is well posed by an argument attributed to Dellachérie (see Rogers-Williams [92], the end of the proof of Theorem 2.5.1). In fact it it were not zero we could take a suitable stopping time τ to ensure $(1 + K_{\tau \wedge t})\mathbf{P}^{\mathcal{L}}_{x_0}$ solves the martingale problem up to time t since

$$K_{\tau \wedge t} M^{df}_s \equiv K_{\tau \wedge t}\left(f(x_s) - f(x_0) - \int_0^s \mathcal{L}f(x_s)ds\right), \qquad 0 \leqslant s \leqslant t$$

is a uniformly integrable martingale.   □

We will often write

$$M^\alpha_t = \int_0^t \alpha_s\,d\{y_s\} \tag{4.5}$$

bringing out the fact that it is the martingale part of the Stratonovitch integral $\int_0^t \alpha_s \circ dy_s$ of (α_t) along the diffusion process (y_t) when that integral is defined, *e.g.* when (α_t) is a continuous semi-martingale. Indeed

Lemma 4.1.2. *Let α be a C^2 1-form, then*

$$M_t^\alpha = \int_0^t \alpha_{y_s} \circ dy_s - \int_0^t (\delta^{\mathcal{L}}\alpha)(y_s)ds, \qquad 0 \leqslant t < \zeta. \qquad (4.6)$$

Proof. This is clear for an exact 1-form. Suppose $\lambda : M \to \mathbf{R}$ is C^2 and α is exact, then for $t < \zeta$,

$$\begin{aligned}
M_t^{\lambda\alpha} &= \int_0^t \lambda(y_s)dM_s^\alpha = \int_0^t \lambda(y_s) \circ dM_s^\alpha - \frac{1}{2}\left\langle \int_0^\cdot d\lambda(y_s)dy_s, M_\cdot^\alpha \right\rangle_t \\
&= \int_0^t \lambda(y_s) \alpha_{y_s} \circ dy_s - \int_0^t \lambda(y_s)(\delta^{\mathcal{L}}\alpha)(y_s)ds - \frac{1}{2}\langle M_\cdot^{d\lambda}, M_\cdot^\alpha \rangle_t \\
&= \int_0^t \lambda(y_s) \alpha_{y_s} \circ dy_s - \int_0^t \delta^{\mathcal{L}}(\lambda\alpha)(y_s)ds
\end{aligned}$$

since $M^{d\lambda}$ is the martingale part of $\lambda(y_s)$ and

$$\langle M^{d\lambda}, M^\alpha \rangle_t = 2 \int_0^t \sigma^{\mathcal{L}}(d\lambda_s, \alpha_s)ds.$$

This proves the result for general α by taking a suitable representation. $\square$

Let S_x be the image of $\sigma_x^{\mathcal{L}}$ in $T_x M$ and let $S := \cup_x S_x$. By a predictable S^*-valued process (α_t) over $(y_t : 0 \leqslant t < \zeta)$ we mean a process $(\alpha_t : 0 \leqslant t)$ such that

(i) $\alpha_t \in S_{y_t}^*$ for all $0 \leqslant t < \zeta$,

(ii) $(\alpha_t \circ \sigma_{y_t}^{\mathcal{L}}, 0 \leqslant t < \zeta)$ is a predictable process in TM, canonically identified with $T^{**}M$.

Note that condition (ii) is equivalent to

(ii)' there exists a predictable $(\bar{\alpha}_t)$ in T^*M over (y_t) such that $\bar{\alpha}_t|_{S_{y_t}} = \alpha_t$ for all $0 \leqslant t < \zeta$.

That (ii') implies (ii) is immediate. To see (ii) implies (ii') first note that $\alpha_t \circ \sigma_{y_t}^{\mathcal{L}} \in S_{y_t}$ for each t since $\alpha_t \circ \sigma_{y_t}^{\mathcal{L}} = \sigma_{y_t}^{\mathcal{L}}(\tilde{\alpha}_t)$ for any extension $\tilde{\alpha}_t$ of α_t to $T_{y_t}^* M$. We can then choose a measurable selection $\bar{\alpha}_t$ in $T_{y_t}^* M$ with $\sigma_{y_t}^{\mathcal{L}}(\bar{\alpha}_t) = \alpha_t \circ \sigma_{y_t}^{\mathcal{L}}$. This process $\bar{\alpha}_t$ will satisfy the requirements of (ii') since

$$\bar{\alpha}_t \sigma_{y_t}^{\mathcal{L}} = \sigma_{y_t}^{\mathcal{L}} \bar{\alpha}_t = \alpha_t \sigma_{y_t}^{\mathcal{L}}. \qquad (4.7)$$

Definition 4.1.3. If (α_t) satisfies (i) and (ii) we will say it is in $L_{\mathcal{L}}^2$ if

$$\int_0^t \alpha_s \sigma_{y_s}^{\mathcal{L}}(\alpha_s)ds < \infty$$

for all $t \geqslant 0$, and will say it is in $L^2_{\mathcal{L},loc}$ if for any compact subset C of M

$$\mathbf{E} \int_0^{t \wedge \zeta} \chi_C(y_s) \alpha_s(\sigma^{\mathcal{L}} \alpha_s) \, ds < \infty$$

for all $t \geqslant 0$.

Remark 4.1.4. Suppose the processes associated to diffusion operators $\mathcal{L}$ and $\mathcal{L} + \mathbf{L}_b$ are both non-explosive, where b is a locally bounded measurable vector field on M. Assume that there exists a T^*M-valued measurable process $b^\#$ defined on the canonical probability space $\mathcal{C}_{y_0} M$ such that $\mathbf{P}^{\mathcal{L}}$-almost surely:

1. $2\sigma^{\mathcal{L}}(b_s^\#) = b(y_s)$,

2. $\int_0^t b_s^\# \sigma^{\mathcal{L}}(b_s^\#) ds < \infty$.

Then, by the GMCM-theorem, as in the Appendix, Section 9.1, we have on $\mathcal{C}([0, T]; M)$,

$$\mathbf{P}^{\mathcal{L} + \mathbf{L}_b} = Z_t \mathbf{P}^{\mathcal{L}}$$

where $Z_t = \exp\{M_t^{b^\#} - \int_0^t b_s^\# \sigma^{\mathcal{L}}(b_s^\#) ds\}$. In an obvious notation, for suitable α, as canonical processes we have, almost surely,

$$\int_0^t \alpha_s d\{y_s\}^{\mathcal{L}} = \int_0^t \alpha_s d\{y_s\}^{\mathcal{L} + \mathbf{L}_b} - \int_0^t \alpha_s(b(u_s)) ds.$$

Lemma 4.1.5. *Suppose $\sigma^{\mathcal{L}}$ has its image in a subset S of TM. Then (M_t^α) depends only on the restriction of α_s in $\mathcal{L}(T_{y_s}M; \mathbf{R})$ to S_{y_s}, $0 \leqslant s < \zeta$. In particular (4.2) defines uniquely a local martingale for each predictable S^*-valued process (α_t) over (y_t) for which the right-hand side of (4.2) is always finite almost surely.*

Proof. For T^*M-valued $\mathcal{F}_*^{y_0}$ predictable processes $(\alpha_t^1, 0 \leqslant t < \zeta)$ and $(\alpha_t^2, 0 \leqslant t < \zeta)$ over $(y_t, 0 \leqslant t < \zeta)$ which agree on S we see

$$\langle M^{\alpha^1} - M^{\alpha^2}, M^{df} \rangle_t = 2 \int_0^{t \wedge \zeta} \sigma(\alpha_s^1 - \alpha_s^2, (df)_{y_s}) \, ds = 0$$

for all $f \in C_c^\infty M$. Therefore $M^{\alpha^1} = M^{\alpha^2}$. On the other hand this also shows that if $\alpha_s \in S_{y_s}^*$ for all s, we can use condition (ii') above to choose a predictable process $\{\bar{\alpha}_s : 0 \leqslant s < \zeta\}$ with values in T^*M over (y_t) and set $M^\alpha = M^{\bar{\alpha}}$ without ambiguity. $\qquad\square$

Example 4.1.6 (Canonical Brownian motion associated to a cohesive diffusion). For simplicity assume that our $\mathcal{L}$-diffusion from a given point y_0 is non-explosive. If $\mathcal{L}$ is cohesive with subbundle E of TM, take a metric connection Γ for E, using the metric determined by $2\sigma^{\mathcal{L}}$. Let

$$\alpha_s(\sigma) := (/\!/_s^\sigma)^{-1} : E_{\sigma(s)} \to E_{y_0}$$

be the inverse of parallel translation, $/\!/_s^\sigma$, along σ from $E_{\sigma(0)}$ to $E_{\sigma(s)}$, for $\mathbf{P}^{y_0}$ almost all paths σ in M. Each component of this with respect to an orthonormal basis for E_{y_0} clearly lies in $L_{\mathcal{L}}^2$. With the obvious extension of our notation to the vector-space-valued case define an E_{y_0}-valued process $B_t : t \geqslant 0$ by

$$B_t = M_t^\alpha = \int_0^t (/\!/_s)^{-1} d\{y_s\}.$$

It is easy to check from its quadratic variation that it is a Brownian motion on the inner product space E_{y_0}. Moreover (as described in [36]) it has the same filtration as the canonical process on $\mathcal{C}_{y_0} M$ up to sets of measure zero. It is the martingale part of the stochastic anti-development $\int_0^t (/\!/_s)^{-1} dy_s$ of our $\mathcal{L}$-diffusion from y_0. The use of a different metric connection would change it by a random rotation, so this process is defined on the canonical probability space $\{\mathcal{C}_{y_0} M, \mathcal{F}^{y_0}, \mathbf{P}^{y_0}\}$ and up to such rotations depends only on it. We have, for α as usual:

$$\int_0^t \alpha_s d\{y_s\} = \int_0^t (\alpha_s \circ /\!/_s) \, dB_s. \tag{4.8}$$

Using the definitions in Appendix 9.3 we see that if our diffusion process $y.$ is a Γ-martingale, then

$$\int_0^t \alpha_s d\{y_s\} = (\Gamma) \int_0^t \alpha_s dy_s. \tag{4.9}$$

Note that there is always some metric connection Γ on E for which a cohesive diffusion process is a Γ-martingale, by section 2.1 of [36].

Example 4.1.7. In Example 2.2.13, the martingales $M_t^{dv_i}$ associated with the process u_t are independent Wiener processes which we will denote by W_t^i, and we have:

$$v_i(u_t) = v_i(u_0) + W_t^i \tag{4.10}$$

$$w_i(u_t) = w_i(u_0) + \alpha \int_0^t \left(v_{i+1}(u_s) dW_s^{i+2} - v_{i+2}(u) dW_s^{i+1} \right).$$

In particular, it follows that:

$$w_3(u_t) = w_3(u_0) + \alpha \int_0^t \left(v_2(u_s) dW_s^3 - v_3(u_s) dW_s^2 \right), \text{ and by Itô's formula,}$$

$$w_1(u_t) = w_1(u_0) + \alpha \left(-v_2(u_t) W_t^3 - v_3(u_0) v_2(u_t) + 2 \int_0^t v_2(u_s) dW_s^3 \right),$$

$$w_2(u_t) = w_2(u_0) + \alpha \left(v_1(u_0) W_t^3 + v_3(u_0) v_1(u_t) - 2 \int_0^t v_1(u_s) dW_s^3 \right).$$

4.2   Horizontality and filtrations

We can characterise horizontality of a diffusion operator or process in terms of filtrations using the following lemma:

Lemma 4.2.1. *Suppose* $p : N \to M$ *is a smooth map,* $\mathcal{B}$ *a smooth diffusion operator over a smooth diffusion operator* $\mathcal{A}$, *and also*

(i) $\sigma^{\mathcal{A}}$ *and* $\sigma^{\mathcal{B}}$ *have constant rank and*

(ii) *the filtration generated by* $u.$ *and* $p(u.)$ *agree up to sets of* $\mathbf{P}^{\mathcal{B}}_{u_0}$*-measure zero for some* $u_0 \in N$.

Then $\operatorname{rank} \sigma^{\mathcal{B}}_u = \operatorname{rank} \sigma^{\mathcal{A}}_{p(u)}$ *for all* $u \in N$.

Proof. Set $p = \operatorname{rank} \sigma^{\mathcal{A}}_x$ and $\tilde{p} = \operatorname{rank} \sigma^{\mathcal{B}}_u$. By assumption p and $\tilde{p}$ do not depend on $x \in M$ and $u \in N$. Take connections on $\operatorname{Image} \sigma^{\mathcal{B}}$ and $\operatorname{Image} \sigma^{\mathcal{A}}$ which are metric for the metrics induced by the symbols. Extend these connections to TN and TM. The martingale part of the stochastic anti-development of $(u.)$ will be a Brownian motion stopped at $\zeta^{\mathcal{B}}$ of dimension $\tilde{p}$ and that of $(p(u.))$ will be one of dimension p. By (ii) these have the same filtration up to sets of measure zero. But this implies $p = \tilde{p}$ by the martingale representation theorem, as required. $\square$

Proposition 4.2.2. *The following are equivalent for* $\mathcal{B}$ *over* $\mathcal{A}$ *when* $\mathcal{A}$ *is cohesive:*

(a) $\mathcal{B} = \mathcal{A}^H$,

(b) $\mathcal{B}$ *is cohesive and the filtration generated by its associated diffusion* $(u.)$ *agrees with that of* $p(u.)$ *up to sets of* $\mathbf{P}^{\mathcal{B}}_{u_0}$*-measure zero for given* u_0 *in* N.

Proof. If (b) holds, Lemma 4.2.1 shows that $\operatorname{Image}[\sigma^{\mathcal{B}}_u] = H_u$ for each $u \in N$, since by (2.3) we always have $H_u \subset \operatorname{Image}[\sigma^{\mathcal{B}}_u]$. Thus (b) implies criterion (ii) of Proposition 2.2.2. Also (b) follows from (iii) of Proposition 2.2.2 by considering the stochastic differential equation driven by horizontal lifts $\tilde{X}^0, \ldots, \tilde{X}^m$. $\square$

4.3   Intertwined diffusion processes

Let $p : N \to M$ be a smooth surjective map. Suppose that $\mathcal{B}$ is over $\mathcal{A}$. However we do not assume $\sigma^{\mathcal{A}}$ of constant rank. Let $\{\mathbf{P}^{\mathcal{B}}_{u_0}\}$ and $\{\mathbf{P}^{\mathcal{A}}_{x_0}\}$ be, respectively, the solutions to the martingale problem for $\mathcal{B}$ and $\mathcal{A}$ on the canonical spaces $\mathcal{C}M^+$ and $\mathcal{C}N^+$. Denote by (u_t) and (x_t) the corresponding canonical processes with explosion time ζ^N and ζ^M respectively. Note that $\zeta^N \leqslant \zeta^M \circ p$ almost surely with respect to $\mathbf{P}^{\mathcal{B}}_{u_0}$. We shall assume that the paths of the diffusion on N do not explode before their projections on M do, more precisely $\zeta^M \circ p = \zeta^N$ almost surely with respect to $\mathbf{P}^{\mathcal{B}}_{u_0}$ for each u_0, equivalently,

- **Assumption S.**

$$\mathcal{C}^p_{u_0} M^+ := \left\{ \sigma : [0, \infty) \to M^+ : \lim_{t \uparrow \zeta^N} p(u_t) = \Delta \text{ when } \zeta^N(u.) < \infty \right\}$$

has full $\mathbf{P}^{\mathcal{B}}_{u_0}$ measure for each $u_0 \in N$.

This assumption holds for all the examples considered earlier. There are two ways it may fail, given the standing assumption that p is surjective. These are exemplified by $p : U \to \mathbf{R}$, the projection $p(x, y) = x$ of certain open sets U of $\mathbf{R}^2$ onto $\mathbf{R}$. If $\mathcal{B}$ is the usual Laplacian on U, with $N = U$, then Assumption S fails if (i) $U = \{(x, y) \in \mathbf{R}^2 : x^2 + y^2 > 1\}$ and if (ii) $U = \{(x, y) \in \mathbf{R}^2 : y < 1\}$.

Denote by the following the filtrations induced by the processes indicated:

$$
\begin{aligned}
\mathcal{F}^{u_0}_t &= \sigma(u_s, 0 \leqslant s \leqslant t), & \mathcal{F}^{u_0} &= \sigma(y_s, 0 \leqslant s < \infty), \\
\mathcal{F}^{x_0}_t &= \sigma(x_s, 0 \leqslant s \leqslant t), & \mathcal{F}^{x_0} &= \sigma(x_s, 0 \leqslant s < \infty), \\
\mathcal{F}^{p(u_0)}_t &= \sigma(p(u_s), 0 \leqslant s \leqslant t), & \mathcal{F}^{p(u_0)} &= \sigma(p(u_s), 0 \leqslant s < \infty).
\end{aligned}
$$

Proposition 4.3.1. *Under Assumption S, $p_*(\mathbf{P}^{\mathcal{B}}_{u_0}) = \mathbf{P}^{\mathcal{A}}_{p(u_0)}$ and $P^{\mathcal{B}}_t(f \circ p) = P^{\mathcal{A}}_t(f \circ p)$ for all $f \in C^{\infty}_c(M)$.*

Proof. If $p(u_0) = x_0$, $f \in C^{\infty}_c(M)$, we only need to show that $M^{df, \mathcal{A}}_t$ is a martingale with respect to $p^*(\mathbf{P}^{\mathcal{B}}_{u_0})$. Using Assumption S,

$$
\begin{aligned}
M^{df, \mathcal{A}}_t(p(u)) &= f(p(u_t)) - f(p(u_0)) - \int_0^t \mathcal{A} f \circ p(u_s))ds \\
&= f(p(u_t)) - f(p(u_0)) - \int_0^t (\mathcal{B}(f \circ p))(u_s)ds \\
&= M^{d(f \circ p), \mathcal{B}}_t
\end{aligned}
$$

is a martingale with respect to $(\mathcal{F}^{u_0}_t)$ and $\mathbf{P}^{\mathcal{B}}_{u_0}$. Take $s \leqslant t$ and let G be a $\mathcal{F}^{x_0}_s$-measurable function. Then

$$
\begin{aligned}
\mathbf{E}^{p_*(\mathbf{P}^{\mathcal{B}}_{u_0})} \left\{ M^{df, \mathcal{A}}_t G \right\} &= \mathbf{E}^{\mathbf{P}^{\mathcal{B}}_{u_0}} \left\{ M^{df, \mathcal{A}}_t(p(u.))G(p(u.)) \right\} \\
&= \mathbf{E}^{\mathbf{P}^{\mathcal{B}}_{u_0}} \left\{ M^{d(f \circ p), \mathcal{B}}_t G \circ p \right\} = \mathbf{E}^{\mathbf{P}^{\mathcal{B}}_{u_0}} \left\{ M^{d(f \circ p), \mathcal{B}}_s G \circ p) \right\} \\
&= \mathbf{E}^{p_*(\mathbf{P}^{\mathcal{B}}_{u_0})} \left\{ M^{df, \mathcal{A}}_s G \right\}
\end{aligned}
$$

and the required result follows from the uniqueness of the martingale problem for $\mathcal{A}$. $\qquad \square$

We will need the following elementary lemma:

Lemma 4.3.2. *Let $(\Omega, \mathcal{F}, \mathcal{F}_t, \mathbf{P}\}$ be a filtered probability space and $\mathcal{G}_*$ a subfiltration of $\mathcal{F}_*$ with the property that for all $s \geqslant 0$,*

$$
\mathbf{E}\{A|\mathcal{G}_s\} = \mathbf{E}\{\mathbf{E}\{A|\mathcal{F}_s\}|\mathcal{G}\}, \qquad \forall A \in \mathcal{F}, \tag{4.11}
$$

where $\mathcal{G} = \vee_s \mathcal{G}_s$. Then

(i) $(\mathbf{E}\{M_t|\mathcal{G}\}, t \geqslant 0)$ is a $\mathcal{G}_*$-martingale whenever $(M_t : t \geqslant 0)$ is an $\mathcal{F}_*$-martingale;

(ii) For all $\mathcal{G}$-measurable and integrable H,

$$\mathbf{E}\{H|\mathcal{F}_s\} = \mathbf{E}\{H|\mathcal{G}_s\};$$

(iii) $\mathbf{E}\{\mathbf{E}\{A|\mathcal{F}_s\}|\mathcal{G}\} = \mathbf{E}\{\mathbf{E}\{A|\mathcal{G}\}|\mathcal{F}_s\}, \qquad \forall A \in \mathcal{F}.$

Proof. For (i) set $N_t = \mathbf{E}\{M_t|\mathcal{G}\}, 0 \leqslant t < \infty$. By (4.11), (N_t) is $\mathcal{G}_t$ measurable. For $s \leqslant t$ suppose that f is $\mathcal{G}_s$-measurable and bounded. Then $\mathbf{E}(N_t f) = \mathbf{E}(M_t f) = \mathbf{E}(M_s f) = \mathbf{E}(N_s f)$. For (ii), let H and F be bounded measurable functions with $\mathcal{G}$-measurable and $\mathcal{F}_s$-measurable representations. Then

$$\mathbf{E}\{H|F\} = \mathbf{E}\{H|\mathbf{E}\{F|\mathcal{G}\}\} = \mathbf{E}\{H|\mathbf{E}\{F|\mathcal{G}_s\}\} = \mathbf{E}\{H|\mathbf{E}\{F|\mathcal{G}\}\}$$

using (4.11). Thus $\mathbf{E}\{H|\mathcal{F}_s\} = \mathbf{E}\{H|\mathcal{G}_s\}$ as required. Part (iii) follows from (ii) on taking $H = \mathbf{E}\{\mathbf{E}\{A|\mathcal{G}\}$ and using equation (4.3.2). □

Part (ii) of the following proposition says that the filtration $\mathcal{F}_*^{p(u_0)}$ is *immersed in* the filtration $\mathcal{F}_*^{u_0}$ in the terminology of Tsirelson [102].

Proposition 4.3.3.   (i) *For fixed $t > 0$ let f be a bounded $\mathcal{F}_t^{u_0}$-measurable function. Then*

$$\mathbf{E}\left\{f|\mathcal{F}_t^{p(u_0)}\right\} = \mathbf{E}\left\{f|\mathcal{F}_t^{p(u_0)}\right\}.$$

(ii) *All $\mathcal{F}_*^{p(u_0)}$ martingales are $\mathcal{F}_*^{u_0}$ martingales. In fact if $f = G \circ p$ for G an integrable functional on $C(M^+)$ with respect to $\mathbf{P}^{\mathcal{A}}$, we have*

$$\mathbf{E}^{u_0}\{f|\mathcal{F}_t^{p(u_0)}\} = \mathbf{E}^{u_0}\{f|\mathcal{F}_t^{u_0}\}.$$

Proof. (i) Write $f = F(u_s : 0 \leqslant s \leqslant t)$ for F a bounded measurable function on CN^+. Let G be bounded measurable functions of $\{p(u_s) : 0 \leqslant s \leqslant t\}$ and $g^1, \ldots, g^k$ bounded Borel functions on M, with $h^1, \ldots, h^k$ positive real numbers. By the Markov property of $u.$ and of $p(u.)$,

$$\mathbf{E}\left(F(u_s : 0 \leqslant s \leqslant t)\, G\, g^1 \circ p(u_{t+h^1}) \cdot \cdots \cdot g^k \circ p(u_{t+h^1+\cdots+h^k})\right)$$
$$= \mathbf{E}\left(F(u_s : 0 \leqslant s \leqslant t)G\, P_{h^1}^{\mathcal{A}}\left(g^1 P_{h^2}^{\mathcal{A}}(g^2 \ldots P_{h^k}^{\mathcal{A}} g^k)\right)(p(u_t))\right).$$

Therefore,

$$\mathbf{E}\left\{F(u_s : 0 \leqslant s \leqslant t)|\mathcal{F}^{p(u)}\right\} = \mathbf{E}\left\{F(u_s : 0 \leqslant s \leqslant t)|\mathcal{F}_t^{p(u)}\right\}$$

as required.

Part (ii) is immediate from (i) by Lemma 4.3.2. □

As in §2.1 set $E_x = \text{Image } \sigma_x^{\mathcal{A}}$ with $h_u : E_{p(u)} \to T_u N$ the horizontal lift defined by (2.3), although now we have no constant rank assumption and so no smoothness of $\mathfrak{h}$. Also let $E_u^{\mathcal{B}} = \text{Image } \sigma_u^{\mathcal{B}}$. For an $\mathcal{F}_*^{x_0}$-predictable E^*-valued process $\phi_t := \phi_t(\sigma.)$, $0 \leqslant t < \zeta^M$ along $(\sigma_t : 0 \leqslant t < \zeta^M)$ let $(p^*(\phi_t) : 0 \leqslant t < \zeta^N)$ be the pull-back restricted to be an $(E^{\mathcal{B}})^*$-valued process along $(u_t : 0 \leqslant t < \zeta^N)$ defined by

$$p^*(\phi_t)(u.) = \phi_t(p(u)) \circ T_{u_t} p : E_{u_t}^{\mathcal{B}} \to \mathbf{R}.$$

Since ϕ_t has a predictable extension $\bar{\phi}_t$ so does $p^*(\phi_t)$ and so the latter is predictable. Moreover $p^*(\phi_t)\sigma^{\mathcal{B}}(p^*\phi_t) = \phi_t\sigma^{\mathcal{A}}(\phi_t)$ by Lemma 2.1.1 showing $\phi.$ is in $L_{\mathcal{A}}^2$ if and only if $p^*(\phi.)$ is in $L_{\mathcal{B}}^2$. For such ϕ we have the following intertwining:

Proposition 4.3.4. *Let ϕ be a predictable $L_{\mathcal{A}}^2$-valued process.*

(1) *For $\mathbf{P}_{u_0}^{\mathcal{B}}$ almost surely all sample paths, $M_t^{\mathcal{A},\phi} \circ p = M_t^{\mathcal{B},p^*(\phi)}$ for $t < \zeta^N$.*

(2) *If $\alpha \in L_{\mathcal{B}}^2$ with $\alpha_t \circ h_{u_t} = 0$ almost surely for all $t < \zeta^N$, then $\langle M_t^{\alpha}, M_t^{df \circ Tp} \rangle = 0$ and $\mathbf{E}^{\mathcal{B},u_0}\{M_t^{\alpha}|\mathcal{F}^{p(u_0)}\} = 0$ for all C^1 functions f on M .*

Proof. For $\phi = df$, (1) follows from $p^*(df)_u = d(f \circ p)_u$ as in the proof of Proposition 4.3.1. For general ϕ, taking a predictable extension if necessary, write $\phi_t(x) = \sum_1^m g_t^j(x.)(df^j)_{x_t}$ for smooth functions $f^j : M \to \mathbf{R}$ and real-valued predictable $\{g_t^j : 0 \leqslant t < \zeta^M\}$. Therefore

$$M_t^{\mathcal{A},\phi} \circ p = \sum_{j=1}^m \int_0^t g_s^j(p(u.)) \, dM_t^{\mathcal{B},p^*(df^j)} = M_t^{\mathcal{B},p^*(\phi)}$$

for all $t < \zeta^N$, giving (1). For (2) let $F : N \to \mathbf{R}$ be a smooth measurable function with respect to $\mathcal{F}^{p(u_0)}$. Then $F = f(p(u.))$ for some measurable function $f : M \to \mathbf{R}$.

$$\mathbf{E}^{\mathcal{B},u_0}\left(M_t^{\alpha} f(p(u.))\right) = \frac{1}{2}\mathbf{E}^{\mathcal{B},u_0}\langle M_t^{\alpha}, M_t^{df \circ Tp}\rangle = \frac{1}{2}\mathbf{E}^{\mathcal{B},u_0}\int_0^t \sigma^{\mathcal{B}}(\alpha_s, df \circ Tp(u_s))ds.$$

If $\alpha_t h_{u_t} = 0$ almost surely for all t, we apply (2.3) to see

$$\sigma^{\mathcal{B}}(\alpha_s, df \circ Tp(u_s)) = \alpha_s \sigma_{u_s}^{\mathcal{B}}\left(T^* p(df)\right) = \alpha_s h_{u_s} \sigma_{p(u_s)}^{\mathcal{A}} df = 0$$

and thus $\mathbf{E}^{\mathcal{B},u_0}(M_t^{\alpha} f(p(u.))) = 0$ giving (2). $\qquad\square$

For $\alpha \in L_{\mathcal{B}}^2$ define $\beta_s \equiv \mathbf{E}^{\mathcal{B},u_0}\{\alpha_s \circ h_{u_s} | p(u.) = x.\}, 0 \leqslant s < \zeta$ to be the unique, up to equivalence, element of $L_{\mathcal{A}}^2$ such that

$$\mathbf{E}^{\mathcal{B},u_0}\left(\alpha_s \circ h_{u_s} \sigma^{\mathcal{A}}(\phi_s(p(u.)))\right) = \mathbf{E}^{\mathcal{A},p(u_0)}\left(\beta_s \sigma^{\mathcal{A}}(\phi_s)\right) \qquad (4.12)$$

for any $\phi \in L_{\mathcal{A}}^2$. To see that such an element exists and is unique, recall that

$$\alpha_s \circ h_{u_s} \sigma_{p(u_s)}^{\mathcal{A}} = \alpha_s \sigma_{u_s}^{\mathcal{B}}(t_{u_s} p)^*$$

which is an $\mathcal{F}_*^{u_0}$-predictable process with values in $E_{p(u_s)} \subset T_{p(u_s)}M$ at each time s, and by Proposition 4.3.3, (4.12) is equivalent to

$$\beta_s(p(u.))\sigma_{p(u_s)}^{\mathcal{A}} = \mathbf{E}^{\mathcal{B},u_0}\left\{\alpha_s\sigma_{u_s}^{\mathcal{B}}(T_{u_s}p)^*|\mathcal{F}^{p(u_0)}\right\} \tag{4.13}$$

in the sense of Elworthy-LeJan-Li [36]. The predictable projection theorem and the results of [36] show that there is a unique, up to indistinguishability, $\mathcal{F}^{p(u.)}$-predictable TM version $\{\gamma_t : 0 \leqslant t < \zeta\}$ say, over $\{p(u_t) : 0 \leqslant t < \zeta\}$, of the right-hand side of (4.13). By applying the uniqueness part of this projection theorem to $\{\phi_s(\gamma_s) : 0 \leqslant s < \zeta\}$ when ϕ. is $\mathcal{F}_*^{p(u.)}$-predictable, T^*M-valued over $p(u.)$ and ϕ_t vanishes on $E_{p(u_t)}$ for all $0 \leqslant t < \rho$ with probability 1, we see $\gamma_t \in E_{p(u_u)}$ for all $0 \leqslant t < \zeta$ almost surely. Now set $\beta_s(p(u.)) = [\sigma_{p(u_s)}^{\mathcal{A}}]^{-1}\gamma_s$ in $E_{p(u_s)}^*$.

Proposition 4.3.5. *For any α. in $L_{\mathcal{B}}^2$ we have*

$$\mathbf{E}^{\mathcal{B},u_0}\{M_t^\alpha \,|\, p(u.) = x.\} = \int_0^T \mathbf{E}^{\mathcal{B},u_0}\{\alpha_s \circ h_{u_s} \,|\, p(u.) = x.\} \, d\{x_s\}.$$

Proof. Set $N_t = \mathbf{E}\{M_t^\alpha \,|\, \mathcal{F}^{p(u_0)}\}$ and write $N_t(u) = \bar{N}_t(p(u))$ for $\{\bar{N}_t\}$ a $\mathcal{F}_t^{x_0}$-measurable function. By Proposition 4.3.3, (N_t) is an $\mathcal{F}_*^{p(u.)}$-martingale and we see $(\bar{N}_t)$ is an $\mathcal{F}_*^x$ martingale. Take $g \in C_c^\infty M$, then by Proposition 4.3.4,

$$\langle \bar{N}, M^{\mathcal{A},dg}\rangle_t \circ p(u) = \mathbf{E}^{\mathcal{B},u_0}\left\{\langle M^\alpha, M^{d(g \circ p)}\rangle_t | \mathcal{F}_t^{p(u_0)}\right\}(u)$$

$$= \mathbf{E}^{\mathcal{B},u_0}\left\{\sigma_{u_t}^{\mathcal{B}}(\alpha_t, (T_{u_t}p)^*(dg)) | \mathcal{F}_t^{p(u_0)}\right\}$$

$$= \mathbf{E}^{\mathcal{B},u_0}\left\{\alpha_t \circ h_{u_t}\sigma_{p(u_t)}^{\mathcal{A}}(dg) | \mathcal{F}_t^{p(u_0)}\right\}$$

by equation (2.3). By Proposition 4.1.1 and the definition above of the conditional expectation, $\bar{N}_t(p(u.)) = M^{\mathcal{A},\beta}$ for $\beta \circ p(u.) = \mathbf{E}^{\mathcal{B},u_0}\{\alpha_t \circ h_{u_t}|\mathcal{F}^{p(u_0)}\}$ and so

$$\bar{N}_t(x.) = \int_0^t \mathbf{E}\{\alpha_s \circ h_{u_s}|p(u.) = x.\} \, d\{x_s\}$$

as required. $\square$

4.4   A family of Markovian kernels

For a probability measure μ_0 on N^+ let the measures μ_t on N^+ be the distributions of u_t, $t \geqslant 0$, under $\mathbf{P}_{\mu_0}^{\mathcal{B}}$ and set $\nu_t = p_*(\mu_t)$ on M^+. Let η_{μ_0} be the law of $u. \mapsto (p(u.), u_0)$ on $CM^+ \times N^+$ under $\mathbf{P}_{\mu_0}^{\mathcal{B}}$ so

$$\eta_{\mu_0}(A,\Gamma) = \int_{y \in M^+} \mathbf{P}_y^{\mathcal{A}}(A)\,\rho_{\mu_0}^y(\Gamma)\,\nu_0(dy), \qquad A \in \mathcal{B}(M^+), \Gamma \in \mathcal{B}(N^+)$$

where $\rho_{\mu_0}^y$ arises from a disintegration of μ_0,

$$\mu_0(\Gamma) = \int_{y \in M^+} \rho_{\mu_0}^y(\Gamma)\,\nu_0(dy), \qquad \Gamma \in \mathcal{B}(N^+).$$

For a measurable $f : N^+ \to \mathbf{R}$, integrable with respect to μ_t set

$$\pi_t^{\mu_0,\sigma} f(v) = \mathbf{E}_{\mu_0}^{\mathcal{B}}\{f(u_t)|p(u.) = \sigma, u_0 = v\}. \tag{4.14}$$

It is defined for η_{μ_0} almost all (σ, v) in $C(M^+) \times N^+$ and depends on the family of μ_0- measure zero sets rather than on μ_0 itself. In particular for $\mathbf{P}_{\nu_0}^{\mathcal{A}}$-almost all σ it is defined for $\rho_{\mu_0}^{\sigma(0)}$-almost all $v \in N^+$. We could use the convention that

$$\pi_t^{\mu_0,\sigma} f(v) = 0$$

if $p(v) \neq \sigma(0)$. This enables us to choose a version of $\mathbf{E}^{\mathcal{B},v}\{f(u_t)|p(u.) = \sigma\}$ which is jointly measurable in σ and v. With this convention, if we define $\theta_t\sigma(s) = \sigma(t+s)$ we see that for $\mathbf{P}_{\nu_0}^{\mathcal{A}}$-almost all σ the map $y \mapsto \pi_t^{\mu_t,\theta_t\sigma} f(y)$ is defined for μ_t-almost all y in N^+.

Further for $u_0 \in N$ and $f : N^+ \to \mathbf{R}$ bounded measurable define

$$\pi_t f(u_0) : \mathcal{C}_{p(u_0)} M^+ \to \mathbf{R},$$

$\mathbf{P}_{p(u_0)}^{\mathcal{A}}$-almost surely, by

$$\pi_t f(u_0)(\sigma) = \mathbf{E}\{f(u_t)|p(u.) = \sigma\} = \pi_t^{\delta_{u_0},\sigma} f(u_0). \tag{4.15}$$

This can be extended, as in [36], to the case of predictable processes in vector bundles over N. In particular if $\alpha.$ is a predictable $\sigma^{\mathcal{B}}[T^*N]^*$-valued process along $u.$, set $h^*(\alpha.)_t = \alpha_t \circ h_{u_t}$. If $\sigma^{\mathcal{A}}$ has constant rank this is a predictable process with values in the pull-back, $p^*(E)$, of E by p. In general it can be considered as an $\mathcal{F}_*^N$-predictable E^*-valued process because

$$(\alpha_t \circ h_{u_t}) \circ \sigma^{\mathcal{A}} = \alpha_t \circ \sigma^{\mathcal{B}} \circ (T_{u_t}p)^* \in T_{x_t}^{**}M.$$

In any case we can define

$$\pi_t(h^*(\alpha.))(u_0) : \mathcal{C}_{p(u_0)} M^+ \to \mathbf{R}$$

as $\mathbf{E}^{\mathcal{B},u_0}\{h^*(\alpha.)_t|p(u.) = x.\}$.

4.5 The filtering equation

Theorem 4.5.1. (1) *If f is $\mathcal{C}_c^2 N$, or more generally if f is C^2 with $\mathcal{B}f$ and $\sigma^{\mathcal{B}}(df, df)$ bounded, then*

$$\pi_t f(u_0) = f(u_0) + \int_0^t \pi_s(\mathcal{B}f)(u_0)ds + \int_0^t \pi_s(df \circ h_{u.})(u_0)d\{x_s\}. \tag{4.16}$$

In particular $\{\pi_t f(u_0) : t \geqslant 0\}$ is a continuous $\mathcal{F}_*^{p(u_0)}$ semi-martingale.

(2) *For bounded measurable $f : M^+ \to \mathbf{R}$ and $\mathbf{P}_{v_0}^{\mathcal{A}}$ almost all σ in $C(M^+)$, for each $s, t \geqslant 0$,*

$$\pi_{t+s}^{\mu_0,\sigma} f(v) = \pi_t^{\mu_0,\sigma} \pi_s^{\theta_t \sigma, \mu_t} f(v) \tag{4.17}$$

for $\rho_{\mu_0}^{\sigma(0)}$ almost all v in N^+.

(3) *Moreover there exists a family of probability measures $Q_v^{\mu_0,\sigma}$ on $C(N^+)$ defining for η_{μ_0}-almost surely all (σ, v) such that if $F : CN^+ \to \mathbf{R}$ is of the form*

$$F(u.) = f_1(u_{t_1}) \dots f_n(u_{t_n})$$

some $0 \leqslant t_1 < t_2 < \cdots t_n$ and bounded measurable $f_j : N^+ \to \mathbf{R}$, $j = 1, 2, \dots, n$, then

$$\int_{u \in CN^+} F(u) Q_v^{\mu_0,\sigma}(du) = \pi_{t_1}^{\mu_0,\sigma}\left(f_1 \pi_{t_2-t_1}^{\mu_{t_1},\theta_{t_1}\sigma}\left(f_2 \dots \pi_{t_n-t_{n-1}}^{\mu_{t_n},\theta_{t_{n-1}}\sigma} f_n\right)\right)(v)$$

$$= \mathbf{E}_{\mu_0}^{\mathcal{B}}\{F(u.)|p(u.) = \sigma, \ u_0 = v\},$$

η_{μ_0}-almost surely in (σ, v).

Proof. (1) By definition of M^{df} we have

$$f(u_t) = f(u_0) + \int_0^t \mathcal{B}f(u_s)ds + M_t^{df}$$

so

$$\pi_t f(u_0) = f(u_0) + \int_0^t \pi_s \mathcal{B}f(u_0)ds + \mathbf{E}\left\{M_t^{df,\mathcal{B}} \,|\, p(u.) = x\right\} \tag{4.18}$$

and part (1) follows from Proposition 4.3.5.

(2) We observed above that the right-hand side of (4.17) is well defined for $\mathbf{P}_{\mu_0}^{\mathcal{A}}$ almost all σ. The equation then follows from the Markov property.

(3) The existence of regular conditional probabilities in our situation implies the existence of the probabilities $\mathbf{Q}_v^{\mu_0,\sigma}$ as required, together with a standard use of the Markov property. $\square$

Remark 4.5.2. A description of the $\mathbf{Q}_v^{\mu_0,\sigma}$ is given in Section 4.8, in the case where $\mathcal{A}$ is cohesive.

Recall that we have the decomposition $F_u = H_u + VT_uN$ for each $u \in N$, and $F = \sqcup F_u$. If $\ell \in F_u^*$ there is a corresponding decomposition

$$\ell = \ell^H + \ell^V \in F_u^*,$$

where ℓ^H vanishes on VT_uN and ℓ^V on H_u. For $\ell \in T_u^*N$ write $\ell^V = (\ell|F_u)^V$ and $\ell^H = (\ell|F_u)^H$.

Corollary 4.5.3. *Suppose $\mathcal{A}$ is cohesive. If f is $C^3_c N$, then there is the Stratonovitch equation*

$$\pi_t f(u_0)(x.) = f(u_0) + \int_0^t \pi_s(\mathcal{B}^V f)(u_0)ds + \int_0^t \pi_s(df_{u_0} \circ h_{u_o}) \circ dx_s. \quad (4.19)$$

Proof. We use (4.18). By Proposition 4.3.5,

$$\mathbf{E}\{M_t^{df} \,|\, p(u.) = x.\} = \mathbf{E}\{M_t^{df^H} \,|\, p(u.) = x.\}.$$

Note that

$$M_t^{df^H} = \int_0^t (df^H)_{u_s} \circ du_s - \int_0^t \delta^{\mathcal{B}}(df^H)(u_s)ds$$

by Lemma 4.1.2. Furthermore

$$\delta^{\mathcal{B}}(df^H) = \delta^{\mathcal{B}^V}(df^H) + \delta^{\mathcal{A}^H}(df^H) = \delta^{\mathcal{A}^H}(df^H) = \delta^{\mathcal{A}^H}(df) = \mathcal{A}^H(f)$$

since df^H vanishes on vertical vectors and $df = df^H + df^V$ while df^V vanishes on horizontal vectors, so $\delta^{\mathcal{A}^H}(df^V) = 0$. This gives

$$\pi_t f(u_0)(x.) = f(u_0) + \int_0^t \pi_s(\mathcal{B}^V f)(u_0)(x.)ds + \mathbf{E}\left\{\int_0^t (df)^H_{u_s} \circ du_s \,\Big|\, p(u.) = x.\right\}$$

Finally (4.19) follows since $df^H_u = p^*(df \circ h_u) = df \circ h_u \circ T_u p$ and $T_u p \circ du_t = \circ dx_t$. $\square$

4.6 Approximations

Assume now that the law of u_t under $\mathbf{P}^{\mathcal{A}}_{u_0}$ is given by

$$P^{\mathcal{A}}_t(u_0, A) = \int_A P^{\mathcal{A}}_t(u_0, v)dv, \qquad A \in \mathcal{B}(M)$$

for $p^{\mathcal{A}}_t(u_0, v)$ a smooth density with respect to some fixed, smooth, strictly positive measure on M to which 'dv' refers. This is the case if $\mathcal{A}$ is hypoelliptic.

Consider the conditional probability

$$q^{u_0,b}_t(V) = \mathbf{P}^{\mathcal{B}}_{u_0}\{u_t \in V | p(u_t) = b\}, \qquad V \in \mathcal{B}(N)$$

defined for $p^{\mathcal{A}}_t(u_0, -)$ almost surely all b in M. There is the disintegration of $p^{\mathcal{B}}_t(u_0, -)$,

$$p^{\mathcal{B}}_t(u_0, V) = \int_{b \in M} q^{u_0,b}_t(V) p^{\mathcal{A}}_t(p(u_0), db),$$

and the formula

$$\mu^{\mu_0,b}_t(V) = \lim_{\epsilon \downarrow 0} [p^{\mathcal{A}}_t(p(u_0), b)]^{-1} \int_V p^{\mathcal{B}}_{t-\epsilon}(u_0, dv) p^{\mathcal{A}}_\epsilon(p(v), b).$$

Take a nested sequence $\{\Pi^\ell\}_{l=1}^\infty$ of partitions of $[0, t]$,

$$\Pi^l = \{0 = t_0^l < t_1^l < \cdots < t_{k_l}^l = t\},$$

say, with union dense in $[0, t]$. For any continuous bounded $f : N^+ \to \mathbf{R}$ there is the following approximation scheme to complete $\pi_t f(u_0)$:

Proposition 4.6.1.

$$\pi_t f(u_0)(\sigma) = \lim_{l\to\infty} \int q_{t_1^l}^{u_0,\sigma(t_1^l)}(dv_1) q_{t_2^l-t_1^l}^{v_1,\sigma(t_2^l)}(dv_2)\ldots q_{t_{l_k}-t_{l_{k-1}}}^{v_{k_l-1},\sigma(t)}(dv_{k_l})f(v_{k_l})$$

$$= \lim_{l\to\infty} \mathbf{E}_{u_0}^{\mathcal{B}}\{f(u_t) \mid p(u_{t_j}^l) = \sigma(t_j^l), \quad 1 \leqslant j \leqslant k_l\}.$$

Proof. The two versions of the right-hand sides are equal before taking limits. For $l = 1, 2, \ldots$, set

$$S^l f(\sigma) = \mathbf{E}_{u_0}\{f(u_t) \mid p(u_{t_j^l}) = \sigma(t_j^l), \quad 1 \leqslant j \leqslant k_l\}.$$

It is defined for $\mathbf{P}_{x_0}^{\mathcal{A}}$-almost all σ in $C(M^+)$, where $x_0 = p(u_0)$. Let Q^l be the σ-algebra on $C(M^+)$ generated by $\sigma \mapsto (\sigma(t_1^l), \ldots, \sigma(t_j^l))$. Directly from the definitions we see

$$\pi_t^l f = \mathbf{E}\{\pi_t(f)(u_0) \mid Q^l\},$$

and so $\{S^l f\}_{l=1}^\infty$ is a Q^*-martingale. it is bounded and so converges $\mathbf{P}_{x_0}^{\mathcal{A}}$-almost surely. Since $\vee_l Q^l$ is the Borel σ-algebra the limit is $\pi_t f(u_0)$ as required. $\square$

4.7   Krylov-Veretennikov Expansion

Suppose $\mathcal{A} = \sum_{j=1}^m \mathbf{L}_{X^j}\mathbf{L}_{X^j} + \mathbf{L}_A$ for smooth vector fields $\{X^j\}_{j=1}^m$ and A. We will now take $\{x_t : 0 \leqslant t < \zeta\}$ to be the solution to the stochastic differential equation

$$dx_t = X(x_t) \circ dB_t + A(x_t)dt, \tag{4.20}$$

with x_0 given, for a Brownian motion B. on $\mathbf{R}^m$, rather than the canonical process. Here $X(x) : \mathbf{R}^m \to T_x M$ is the map given by

$$X(x)(a^1, \ldots, a^m) = \sum_{j=1}^m a^j X^j(x), \quad x \in M.$$

Let $\{P_t : t \geqslant 0\}$ be the sub-Markovian semi-group generated by $\mathcal{B}$. Let $f \in C_c^\infty N$. Assume $P_t f \in C^\infty N$.

As in the proof of Theorem 4.5.1, from

$$P_{t-s}f(u_s) = P_t f(u_0) + \int_0^s d(P_{t-r}f)_{u_r} d\{u_r\}, \quad 0 \leqslant s \leqslant t$$

we obtain

$$\pi_s P_{t-s}(f)(u_0)(x.) = P_t f(u_0) + \int_0^s \mathbf{E}\{d(P_{t-r}f)_{u_r} \circ h_{u_r} \mid p(u.) = x.\} d\{x_r\}$$

$$= P_t f(u_0) + \int_0^s \mathbf{E}\{d(P_{t-r}f)_{u_r} \circ h_{u_r} \mid p(u.) = x.\} X(x_r) dB_r$$

so that $\pi_s P_{t-s} f(u_0), 0 \leqslant s \leqslant t$, is a continuous $\mathcal{F}_*^{x_0}$ semi-martingale. Therefore

$$\pi_t f(u_0) - P_t f(u_0) = \int_0^t d_s(\pi_s P_{t-s} f(u_s))$$

$$= \int_0^t \mathbf{E}\{d(P_{t-r}f) \circ h_{u_r} \mid p(u.) = x.\} X(x_r) dB_r$$

$$= \int_0^t S_r[d(P_{t-r}f) \circ h_- \circ X^k(p(-))](u_0) dB_r^k$$

giving a 'Clark-Ocone' formula for $\pi_t f(u_0)$. Iterating this procedure formally,

$$\pi_t f(u_0) = P_t f(u_0) + \int_0^t S_r[d(P_{t-r}f) \circ h_- \circ X(p(-))](u_0) dB_r$$

$$+ \int_0^t \int_0^r \pi_s [dP_{r-s}[d(P_{t-r}f) \circ h_- \circ X^k(p(-))]h_- \circ X^j(p(-)] dB_s^j dB_r^k$$

$$= \dots,$$

we obtain the Wiener chaos expansion of $\pi_t f(u_0)(x.)$.

4.8 Conditional Laws

It will be convenient to extend the notation of section 4.3. For $0 \leqslant l < r < \infty$ let $\mathcal{C}(l, r; N^+)$ and $\mathcal{C}(l, r; M^+)$ be respectively the space of continuous paths $u :$ $[l, r] \to N^+$ and $x : [l, r] \to M^+$ which remain at Δ from the time of explosion; and $\mathcal{C}_{u_0}(l, r; N^+)$ and $\mathcal{C}_{x_0}(l, r; M^+)$ the paths from $u_0 \in N^+$ and $x_0 \in M^+$ respectively, Let $\{\mathbf{P}_{u_0}^{(l,r),\mathcal{B}}\}$ and $\{\mathbf{P}_{x_0}^{(l,r),\mathcal{A}}\}$ be the associated diffusion measures.

The conditional law of $\{u_s : l \leqslant s \leqslant r\}$ given $\{p(u_s) : l \leqslant s \leqslant r\}$ will be given by probability kernels $\sigma \mapsto \mathbf{Q}_{\sigma,u_0}^{l,r}$ defined $\mathbf{P}^{(l,r);\mathcal{A}}$ almost surely from $\mathcal{C}_{p(u_0)}(l, r; M^+)$ to $\mathcal{C}_{u_0}^p(l, r; N^+)$ for each $u_0 \in N$, where $\mathcal{C}_{u_0}^p(l, r; N^+)$ is the subspace of $\mathcal{C}_{u_0}(l, r; N^+)$ whose paths satisfy Assumption S. The defining property is that, for integrable $f : \mathcal{C}_{u_0}(l, r; N^+) \to \mathbf{R}$,

$$\mathbf{E}\{f(u.) \mid p(u_s) = \sigma_s, l \leqslant s \leqslant r\} = \int_{y \in \mathcal{C}_{u_0}(l,r;N^+)} f(y) d\mathbf{Q}_{\sigma,u_0}^{l,r}(y). \tag{4.21}$$

To obtain the conditional law take the decomposition $\mathcal{B} = \mathcal{A}^H + \mathcal{B}^V$ of Proposition 2.2.5. Represent the diffusion corresponding to $\mathcal{A}$ by a stochastic differential equation

$$dx'_t = X(x'_t) \circ dB_t + X^0(x'_t)dt. \tag{4.22}$$

Take a connection ∇^V on VTN and let

$$(\nabla^V) \qquad dz_t = V(z_t)dW_t + V^0(z_t)dt \tag{4.23}$$

be an Itô equation whose solutions are $\mathcal{B}^V$-diffusions. Here (W_t) is the canonical Brownian motion on $\mathbf{R}^m$ for some m, independent of $(B.)$, the map $V : M \times \mathbf{R}^m \to TM$ takes values in $\ker[Tp]$, and V and V^0 are locally Lipschitz. For such a representation of $\mathcal{B}^V$ diffusions see the Appendix, Section 9.2.4. Let $\tilde{X} : N \times \mathbf{R}^m \to H$ and $\tilde{X}^0 : N \to H$ be the horizontal lifts of X and X^0 respectively using Theorem 2.1.2. The solution to the equation of mixed type

$$(\nabla^V) \qquad dy_t = \tilde{X}(y_t) \circ dB_t + \tilde{X}^0(y_t)dt + V(y_t)dW_t + V^0(y_t)dt,$$
$$y_l = u_0, \qquad u_0 \in N, \quad l \leqslant t \leqslant r.$$

has law $\mathbf{P}_{u_0}^{(l,r),\mathcal{B}}$. Noting that $\tilde{X}(u) = h_u X(p(u))$ for $u \in M$,

$$(\nabla^V) \qquad dy_t = h_{y_t} \circ dx'_t + V(y_t)dW_t + V^0(y_t)dt, \tag{4.24}$$
$$y_l = u_0, \quad l \leqslant t \leqslant r,$$

where $x'_t = p(y_t)$ so that (x'_t) is a solution to (4.22) starting from $p(u_0)$ at time l. Without changing the law of y. we can replace x' by the canonical process $x..$ Then

Theorem 4.8.1. *Consider the solution (y_t) as a process defined on the probability space $\mathcal{C}_{p(u_0)}(l, r; M^+) \times \mathcal{C}_0\mathbf{R}^m$ with product measure,*

$$y : [l, r] \times \mathcal{C}_{p(u_0)}(l, r; M^+) \times \mathcal{C}_0\mathbf{R}^m \to N^+,$$

and define $\mathbf{Q}_{\sigma, u_0}^{l, r}$ to be the law of $y(\sigma, -) : \mathcal{C}_0\mathbf{R}^m \to \mathcal{C}_{u_0}(l, r; N^+)$. For bounded measurable $f : \mathcal{C}_{u_0}(l, r; N^+) \to \mathbf{R}$,

$$\mathbf{E}\{f(u.) \mid p(u_s) = \sigma_s, l \leqslant s \leqslant r\} = \int_{y \in \mathcal{C}_{u_0}(l,r;N^+)} f(y)d\mathbf{Q}_{\sigma, u_0}^{l,r}(y).$$

Proof. Take a measurable function $\alpha : C_{p(u_0)}(l,r;M^+) \to \mathbf{R}$. Then

$$
\mathbf{E}^{\mathbf{P}^{\mathcal{B}}_{u_0}} \left(\alpha(p(u)) \int_{y \in C_{u_0}(l,r;N^+)} f(y) \, d\mathbf{Q}^{l,r}_{p(u),u_0}(y) \right)
$$

$$
= \mathbf{E}^{\mathbf{P}^{\mathcal{A}}_{p(u_0)}} \left(\alpha(x) \int_{y \in C_{u_0}(l,r;N^+)} f(y) \, d\mathbf{Q}^{l,r}_{x,u_0}(y) \right)
$$

$$
= \mathbf{E}^{\mathbf{P}^{\mathcal{A}}_{p(u_0)}} \left(\alpha(x) \int_{C_0 \mathbf{R}^m} f(y(x,\omega)) \, d\mathbf{P}(\omega) \right)
$$

$$
= \int_{C_{p(u_0)}(l,r;M^+) \times C_0 \mathbf{R}^m} \left(\alpha(x) f(y(x,\omega)) \, d\mathbf{P}^{\mathcal{A}}_{p(u_0)} \, d\mathbf{P}(\omega) \right)
$$

$$
= \mathbf{E} f(u) \alpha(p(u)),
$$

as required. □

Note that Theorem 4.8.1 is equivalent to the statement that $\omega \mapsto \mathbf{Q}^{l,r}_{p(\omega),u_0}$, $\omega \in C_{u_0}(l,r;N^+)$, is a regular conditional probability of $\mathbf{P}^{(l,r),\mathcal{B}}_{u_0}$ given p.

Remark 4.8.2. Let $(\xi^l_t(\cdot,\cdot), l \leqslant t < \infty)$ be a measurable flow for (4.22) and $(\eta^l_t(\sigma,\cdot,), 0 \leqslant t < \infty)$ one for (4.24) with x' replaced by $\sigma \in C_{p(u_0)}(l,r;M^+)$. For $\omega \in \Omega$, the underlying probability space for the Brownian motion $\mathcal{B}$, define $\mathbf{Q}^{l,r}_\omega$, from the space of bounded measurable functions on N^+ to itself, by

$$
\mathbf{Q}^{l,r}_\omega(f)(u_0) = \mathbf{E} f \left(\eta^l_r(\xi^l_r(p(u_0),\omega), u_0) \right).
$$

A direct calculation shows that

$$
\mathbf{Q}^{l,r}_\omega \mathbf{Q}^{r,s}_\omega = \mathbf{Q}^{l,s}_\omega
$$

for $0 \leqslant l \leqslant r \leqslant s < \infty$. Thus their adjoints on a suitable dual space would form an evolution.

More generally, letting Borel(X) stand for the Borel σ-algebra of a topological space X:

Proposition 4.8.3. *Let φ be a measurable map from $C_{x_0}(l,r;M^+)$ to some measure space, and let*

$$
\mathbf{P}^{(l,r),\varphi}_{x_0} : C_{x_0}(l,r;M^+) \times \mathrm{Borel}(C_{x_0}(l,r;M^+)) \to [0,1]
$$

be a regular conditional probability for $\mathbf{P}^{(l,r)}_{x_0}$ given φ. For u_0 with $p(u_0) = x_0$ set

$$
Q^{l,r,\varphi \circ p}_{u_0}(\omega, A) = \int_{C_{x_0}(l,r;M^+)} \mathbf{Q}^{l,r}_{\sigma,u_0}(A) \mathbf{P}^{(l,r),\varphi}_{x_0}(p(\omega), d\sigma)
$$

for $\omega \in C_{u_0}(l,r;N^+)$ and $A \in \mathrm{Borel}(C_{u_0}(l,r;N^+))$. Then $Q^{l,r,\varphi \circ p}_{u_0}$ is a regular conditional probability of $\mathbf{P}^{(l,r),\mathcal{B}}_{u_0}$ given $\varphi \circ p$.

Proof. By definition

$$
\mathbf{Q}_{u_0}^{l,r,\varphi \circ p}(\omega, A) = \mathbf{E}^{(l,r),\mathcal{A},x_0} \left\{ \mathbf{Q}_{p(-),u_0}^{l,r}(A) | \varphi \right\} p(\omega)
$$

$$
= \mathbf{E}^{(l,r),\mathcal{A},x_0} \left\{ \mathbf{E}^{(l,r),\mathcal{B},u_0} \{ \chi_A | p = - \} | \varphi \right\} p(\omega)
$$

$$
= \mathbf{E}^{(l,r),\mathcal{B},u_0} \{ \chi_A | \varphi \circ p \}(\omega). \qquad \square
$$

Corollary 4.8.4. *For φ as in Theorem 4.8.3 suppose that the canonical process on M^+ with law $\mathbf{P}_{x_0}^{(0,T),\varphi}(\sigma, -)$ is a semi-martingale for almost all σ, in its own filtration $\mathcal{F}_t^{x_0}, 0 \leqslant t \leqslant T$, for $\mathbf{P}_{x_0}^{(0,T),\mathcal{A}}$ almost all σ. Then the solution $y(\sigma, -)$ to the equation*

$$
(\nabla^V) \qquad dy_t = h_{y_t} \circ d\sigma_t + V(y_t)dW_t + V^0(y_t)dt, \qquad (4.25)
$$
$$
y_l = u_0, \quad 0 \leqslant t \leqslant T
$$

where $\sigma_t, 0 \leqslant t \leqslant T$ is run with law $\mathbf{P}_{x_0}^{(0,T),\varphi}(\sigma, -)$, is a version of the $\mathcal{B}$-diffusion from u_0 conditioned by $\varphi \circ p$.

Proof. That the law of the solution is as required follows from the discussion at the beginning of this section together with Proposition 4.8.3 and Fubini's theorem. $\quad \square$

Conditions under which conditioned processes are semi-martingales are discussed by Baudoin [4]. In particular bridge processes derived from elliptic diffusions are, so we obtain the following version of Carverhill's result [16]:

Corollary 4.8.5. *Suppose $\mathcal{A}$ is elliptic and let $b_t : 0 \leqslant t \leqslant T$ be a version of the $\mathcal{A}$-bridge going from x_0 to z in time T, some $z \in M$. Then the solutions to*

$$
(\nabla^V) \qquad dy_t = h_{y_t} \circ db_t + V(y_t)dW_t + V^0(y_t)dt, \qquad (4.26)
$$
$$
y_0 = u_0, \quad 0 \leqslant t \leqslant T
$$

give a version of the $\mathcal{B}$ diffusion from u_0 conditioned on $p(u_T) = z$.

Example 4.8.6. In Example 2.2.13 and its continuation Example 4.1.7, for all $t > 0$ the conditional distribution of $(u_s, s \leq t)$ given u_0 and the path $(p(u_s), s \leq t)$ is determined by the conditional distribution of $((v_3, w_1, w_2)(u_s), s \leq t)$ given $\sigma(W_s^1, W_s^2, s \leq t)$ (Cf formula (4.10) for the definition of the processes W^i). It is the law of

$$
\Bigg((v_3(u_0) + W_t, w_1(u_0) - v_2(u_0)W_t - v_3(u_0)v_2(u_t) + 2 \int_0^t v_2(u_s)dW_s,
$$
$$
w_2(u_0) + v_1(u_0)W_t + v_3(u_0)v_1(u_t) - 2 \int_0^t v_1(u_s)dW_s \Bigg),
$$

where W_t is a Wiener process independent of W^1 and W^2. In particular, the conditional distribution of u_t given u_0 and the path $(p(u_s), s \leq t)$ is a Gaussian distribution with mean vector

$$\big(v_3(u_0), w_1(u_0) - v_3(u_0)v_2(u_t), w_2(u_0) + v_3(u_0)v_1(u_t)\big).$$

Set $g(s) = (1, 2v_2(u_s) - v_2(u_0), v_1(u_0) - 2v_1(u_s))$, then the covariance is given by

$$\int_0^t g(s)g^T(s)\, ds\,.$$

4.9   An SPDE example

Consider now the SPDE

$$du_t(\theta) = \Delta u_t(\theta) + u_t(\theta)dW_t + \phi(\theta)dB_t \tag{4.27}$$

on the circle S^1 where W and B are independent one-dimensional Brownian motions. Suppose $\phi : S^1 \to \mathbf{R}$ is C^∞. Let N be the Sobolev space $H^s(S^1; \mathbf{R})$, for sufficiently large s. Then for a smooth initial condition the solution to equation (4.27) will have $u_t \in N$ for $t \geq 0$. Take $M = \mathbf{R}$ and define $p : N \to M$ by $p(u) = \int_{S^1} u(\theta)d\theta$. Set $m = \int_{S^1} \phi(\theta)d\theta$.

The process $x_t := p(u_t)$ satisfies

$$dx_t = d\{x_t\} = x_t dW_t + m dB_t$$

and is a diffusion process with generator $\mathcal{A} = \frac{1}{2}(x^2 + m^2)\frac{\partial^2}{\partial x^2}$.

If m vanishes, $W_t = \log(\frac{x_t}{x_0}) + \frac{t}{2}$ and conditionally to $x_s, 0 \leq s \leq t$, u_t is a Gaussian Markov process whose law is easily determined.

When m does not vanish, equation (4.16) applies. As an example, for $z \in S^1$, to find the conditional expectation of $u_t(z)$ given $\{\int_{S^1} u_s(\theta)\, d\theta : 0 \leq \theta \leq t\}$ take $f : N \to \mathbf{R}$ to be the evaluation δ_z at z so $f(u) = u(z) = \delta_z(u)$. Then $\mathcal{B}(\delta_z)(u) = \Delta u(z)$ and therefore $\mathcal{B}(\delta_z) = \delta_z \circ \Delta : N \to \mathbf{R}$. Also $(d\delta_z)_u = \delta_z$ for all u since δ_z is linear.

Moreover, using Lemma 2.3.1, we see $h_u(1) = \frac{p(u)u + m\phi}{p(u)^2 + m^2} \in N$ and so $d\delta_z \circ h_u : \mathbf{R} \to \mathbf{R}$ is multiplication by $\frac{p(u)u(z) + m\phi(z)}{p(u)^2 + m^2}$, i.e.

$$(d\delta_z \circ h_u)(\lambda) = \lambda\frac{p\delta_z(u) + m\phi(z)}{p^2 + m^2}, \lambda \in \mathbf{R}.$$

Thus in equation (4.16)

$$\pi_s(d\delta_z \circ h)(\lambda) = \frac{p\pi_s(\delta_z) + m\phi(z)}{p^2 + m^2}(\lambda) \quad \text{and} \quad \pi_s(\mathcal{B}\delta_z) = \pi_s(\delta_z \circ \Delta)$$

and equation (4.16) yields:

$$\pi_t(\delta_z)(u_0) = u_0(z) + \int_0^t \pi_s(\delta_z \circ \Delta)(u_0)ds + \int_0^t \frac{p\pi_s(\delta_z) + m\phi(z)}{p^2 + m^2}(x_s dW_s + m dB_s).$$

$$(4.28)$$

The SDE for the conditional process can be derived using the decomposition of the noise as in Section 2.3.2. For this write

$$du_t = h_{u_t}(1)dx_t + du_t - h_{u_t}(1)dx_t = \frac{x_t u_t + m\phi}{x_t^2 + m^2}dx_t + \Delta u_t dt + dM_t$$

where

$$dM_t = u_t dW_t - \frac{x_t^2 u_t + m x_t \phi}{x_t^2 + m^2}dW_t + \phi dB_t - \frac{x_t u_t m + m^2 \phi}{x_t^2 + m^2}dB_t$$

$$= \frac{m^2 u_t - m x_t \phi}{x_t^2 + m^2}dW_t + \frac{-x_t u_t m + x_t^2 \phi}{x_t^2 + m^2}dB_t.$$

Note that as expected the bracket $\langle M, x \rangle_t$ vanishes. In fact M_t can be obtained by parallel translation of a Brownian motion β_t, independent of x_t, as explained in Section 2.3.2. The conditional expectation of u_t given $\sigma(x_s, 0 \le s \le t)$ is easily obtained by taking its expectation against any stochastic integral with respect to x (which happens to be a Brownian martingale). We see that the two last terms do not contribute so that, if z_t denotes $\mathbb{E}\{u_t | \sigma(x_s, 0 \le s \le t)\}$, we have:

$$dz_t = \frac{x_t z_t + m\phi}{x_t^2 + m^2}dx_t + \Delta z_t dt. \qquad (4.29)$$

since $\pi_s(\delta_z)(u_0) = z_s$, in this case the use of the conditioned process is much simpler than the use of the filtering equation (4.16). The latter yields equation (4.28) which would traditionally be interpreted as an equation for a process with values in the space of measures on our Sobolev space H^s. The former gives the rather straightforward linear SPDE (4.29).

However note that the Fourier coefficients of u_t verify linear SDE's:

$$d\widehat{u}_t(k) = \frac{x_t \widehat{u}_t(k) + m\widehat{\phi}(k)}{x_t^2 + m^2}dx_t - k^2 \widehat{u}_t(k)dt + \frac{m^2 \widehat{u}_t(k) - m x_t \widehat{\phi}(k)}{x_t^2 + m^2}dW_t$$

$$+ \frac{-x_t \widehat{u}_t(k)m + x_t^2 \widehat{\phi}(k)}{x_t^2 + m^2}dB_t.$$

As before we see that the two last terms do not contribute under our conditioning, and for z_t as above,

$$d\widehat{z}_t(k) = \frac{x_t \widehat{z}_t(k) + m\widehat{\phi}(k)}{x_t^2 + m^2}dx_t - k^2 \widehat{z}_t(k)dt.$$

which is a linear equation from which $\hat{z}_t$, and so z_t, can be computed in the standard way.

This formula can be recovered easily from equation (4.16) as, if we take $f(u) = \hat{u}(k)$, we have $\mathcal{B}f = -k^2 f$ and $df \circ h_u = \frac{p(u)f(u)+m\hat{\phi}(k)}{p(u)^2+m^2}$, so that $\pi_s(\mathcal{B}f) = -k^2\pi_s(f)$ and $\pi_s(df \circ h) = \frac{p\pi_s(f)+m\hat{\phi}(k)}{p^2+m^2}$.

More complicated equations can be given for the conditional expectations of functions of $\hat{u}_t(k)$. They involve the vertical part of the generator $\mathcal{B}$.

4.10 Equivariant case: skew-product decomposition

In the equivariant case, when N is the total space P of a principal bundle $\pi : P \to M$ as in Chapter 3, a version of Theorem 4.8.1 is given in [34] which reflects the additional structure. In particular the following is proved there:

Proposition 4.10.1. *Let $\mathcal{B}$ be an equivariant diffusion operator on P which induces a cohesive diffusion operator $\mathcal{A}$ on M. Let $\{y_t : 0 \leqslant t < \zeta\}$ be a $\mathcal{B}$-diffusion on P. Then*

$$y_t = \tilde{x}_t \cdot g_t^{\tilde{x}.},$$

where

(i) $\{\tilde{x}_t : 0 \leqslant t < \zeta\}$ *is the horizontal lift of $p(y.)$, starting at y_0, using the semi-connection induced by $\mathcal{B}$,*

(ii) $\{g_t^\sigma : 0 \leqslant t < \zeta(\sigma)\}$ *is a diffusion independent of $\{p(y_t) : 0 \leqslant t < \zeta\}$ on G starting at the identity with time dependent generator $\mathcal{L}_t^\sigma$ given by*

$$\mathcal{L}_t^\sigma f(g) = \sum_{i,j} \alpha^{ij}(\sigma(t) \cdot g)\mathbf{L}_{A_i^*}\mathbf{L}_{A_j^*} f(g) + \sum \beta^k(\sigma(t)g)\mathbf{L}_{A_k^*} f(g),$$

$0 \leqslant t < \zeta(\sigma)$, *for any $\sigma \in C M^+$, where $A_1^*, \ldots, A_k^*$ are the left invariant vector fields on G corresponding to a basis of $\mathfrak{g}$ and the α^{ij} and β^k are the coefficients for $\mathcal{B}^V$ as in Theorem 3.2.1.*

Note that for each t the operator $\mathcal{L}_t^\sigma$ is conjugate to the restriction of $\mathcal{B}^V$ to the fibre through $\sigma(t)$ by the map

$$G \mapsto p^{-1}(p(\sigma(t))),$$
$$g \mapsto \sigma(t)g.$$

It is a right-invariant operator.

Remark 4.10.2. Note that by the equivariance of $\mathcal{L}^\sigma$ there will be no explosion of the process (g_t^σ) before that of $\sigma..$ Consequently Assumption S of §4.3 holds automatically.

Below we give the equivariant version of Proposition 4.8.1. We shall use the notation of §4.8. However we replace the one-point compactification P^+ of P by $\bar{P} = P \cup \Delta$ with the smallest topology agreeing with that of P and such that $\pi : \bar{P} \to M^+$ is continuous. Also let G^+ be the one-point compactification $G \cup \Delta$ of G with group multiplication and action of G extended so that

$$u \cdot \Delta = \Delta, \Delta \cdot g = g \cdot \Delta = \Delta, \qquad \forall u \in \bar{P}, g \in \bar{G}.$$

For $0 \leqslant l < r < \infty$ if $y \in \mathcal{C}(l, r; \bar{P})$, we write $l_y = l$ and $r_y = r$. Let $\mathcal{C}(*, *; \bar{P})$ be the union of such spaces $\mathcal{C}(l, r; \bar{P})$. It has the standard additive structure under concatenation: if y and y' are two paths with $r_y = l_{y'}$ and $y(r_y) = y'(l_{y'})$ let $y + y'$ be the corresponding element in $C(l_y, r_{y'}; \bar{P})$. The *basic* σ-algebra of $C(*, *, \bar{P})$ is defined to be the pull-back by π of the usual Borel σ-algebra on $C(*, *; M^+)$.

Given an equivariant diffusion operator $\mathcal{B}$ on P consider the laws $\{\mathbf{P}_a^{(l,r),\mathcal{B}} : a \in P\}$ as a kernel from P to $\mathcal{C}(l, r; \bar{P})$. The right action R_g by g in G^+ extends to give a right action, also written R_g, of G^+ on $\mathcal{C}(*, *, \bar{P})$. Equivariance of $\mathcal{B}$ is equivalent to

$$\mathbf{P}_{ag}^{(l,r),\mathcal{B}} = (R_g)_* \mathbf{P}_a^{(l,r),\mathcal{B}}$$

for all $0 \leqslant l \leqslant r$ and $a \in P$. Therefore $\pi_*(\mathbf{P}_a^{(l,r),\mathcal{B}})$ depends only on $\pi(a)$, l, r and gives the law of the induced diffusion $\mathcal{A}$ on M. We say that such a diffusion $\mathcal{B}$ is *basic* if for all $a \in P$ and $0 \leqslant l < r < \infty$ the basic σ-algebra on $\mathcal{C}(l, r; \bar{P})$ contains all Borel sets up to $\mathbf{P}_a^{(l,r),\mathcal{B}}$ negligible sets, i.e. for all $a \in P$ and Borel subsets B of $\mathcal{C}(l, r; \bar{P})$ there exists a Borel subset A of $\mathcal{C}(l, r, M^+)$ s.t. $\mathbf{P}_a^{(l,r),\mathcal{B}}(\pi^{-1}(A) \Delta B) = 0$.

For paths in G it is more convenient to consider the space $\tilde{\mathcal{C}}_{id}(l, r; G^+)$ of cadlag paths $\sigma : [l, r] \to G^+$ with $\sigma(l) = \text{id}$ such that σ is continuous until it leaves G and stays at Δ from then on. It has a multiplication

$$\tilde{\mathcal{C}}_{id}(s, t; G^+) \times \tilde{\mathcal{C}}_{id}(t, u; G^+) \longrightarrow \tilde{\mathcal{C}}_{id}(s, u; G^+)$$

$$(g, g') \mapsto g \times g'$$

where $(g \times g')(r) = g(r)$ for $r \in [s, t]$ and $(g \times g')(r) = g(t)g'(r)$ for $r \in [t, u]$.

Given probability measures $\mathbf{Q}$, $\mathbf{Q}'$ on $\tilde{\mathcal{C}}_{id}(s, t; G^+)$ and $\tilde{\mathcal{C}}_{id}(t, u; G^+)$ respectively this determines a convolution $\mathbf{Q} * \mathbf{Q}'$ of $\mathbf{Q}$ with $\mathbf{Q}'$ which is a probability measure on $\tilde{\mathcal{C}}_{id}(s, u; G^+)$.

Theorem 4.10.3. *Given the laws* $\{\mathbf{P}_a^{(l,r),\mathcal{B}} : a \in P, 0 \leqslant l < r < \infty\}$ *of an equivariant diffusion* $\mathcal{B}$ *over a cohesive* $\mathcal{A}$ *there exist probability kernels* $\{\mathbf{P}_a^{H,l,r} : a \in P\}$ *from* P *to* $\mathcal{C}(l, r; \bar{P})$, $0 \leqslant l < r < \infty$ *and* $y \mapsto \mathbf{Q}_y^{l,r}$, *defined* $\mathbf{P}^{l,r}$ *a.s. from* $\mathcal{C}(l, r; \bar{P})$ *to* $\tilde{\mathcal{C}}_{id}(l, r; G^+)$ *such that*

(i) $\{\mathbf{P}_a^{H,l,r} : a \in P\}$ *is equivariant, basic and determining a cohesive generator.*

(ii) $y \mapsto \mathbf{Q}_y^{l,r}$ *satisfies*

$$\mathbf{Q}_{y+y'}^{l_y, r_{y'}} = \mathbf{Q}_y^{l_y, r_y} * \mathbf{Q}_{y'}^{l_{y'}, r_{y'}}$$

for $\mathbf{P}^{l_y, r_y} \otimes \mathbf{P}^{l_{y'}, r_{y'}}$ *almost all* y, y' *with* $r_y = l_{y'}$.

(iii) *For U a Borel subset of $\mathcal{C}(l, r; \bar{P})$,*

$$\mathbf{P}_a^{l,r}(U) = \int_{\mathcal{C}(l,r;\bar{P})} \int_{\tilde{\mathcal{C}}(l,r;G^+)} \chi_U(y. \cdot g.) \mathbf{Q}_y^{l,r}(dg) \mathbf{P}_a^{H,l,r}(dy).$$

The kernels $\mathbf{P}_a^{H,l,r}$ are uniquely determined as are the $\{\mathbf{Q}_y^{l,r} : y \in \mathcal{C}(l, r; \bar{P})\}$, $\mathbf{P}_a^{H,l,r}$ a.s. in y for all a in P. Furthermore $\mathbf{Q}_y^{l,r}$ depends on y only through its projection $\pi(y)$ and its initial point y_l.

The proof of this theorem is as that of Theorem 2.5 in [34] (although there the processes are assumed to have no explosion).

Stochastic differential equations can be given for $(\tilde{x}_t)$ and (g_t^σ) as in §4.8, from which the decomposition can be proved via Itô's formula; see Theorem 8.2.5 below for details of a special case.

Proposition 4.10.1 extends results for Riemannian submersions by Elworthy-Kendall [33] and related results by Liao[69]. A rich supply of examples of skew-product decomposition of Brownian motions, with a general discussion, is given in Pauwels-Rogers[87].

For a special class of derivative flows, considered as a *GLM*-valued process as in §3.3, there is a different decomposition by Liao [70], see also Ruffino [94].

4.11   Conditional expectations of induced processes on vector bundles

In the notation of §3.4 let $\rho : G \to L(V, V)$ be a C^∞ representation with $\Pi^\rho : F \to M$ the associated bundle. A $\mathcal{B}$-diffusion $\{y_t : 0 \leqslant t < \zeta\}$ on P determines a family of $\{\Psi_t : 0 \leqslant t < \zeta\}$ of random linear maps W_t from $F_{x_0} \to F_{x_t}$, where $x_t = \pi(y_t)$. By definition,

$$\Psi_t[(y_0, e)] = [(y_t, e)].$$

Assuming $\mathcal{A}$ is cohesive we have the parallel translation $/\!/_t : F_{x_0} \to F_{x_t}$ along $\{x_t : 0 \leqslant t < \zeta\}$ determined by our semi-connection. This is given by

$$/\!/_t[(y_0, e)] = [(\tilde{x}_t, e)]$$

where $\tilde{x}.$ is the horizontal lift of $x.$ starting at y_0.

When taken together with Corollary 3.4.9 the following extends results for derivative flows in Elworthy-Yor[38], Li[68], Elworthy-Rosenberg [37], and Elworthy-LeJan-Li[36]. See the Notes below, Section 4.12.

Theorem 4.11.1. *Let $\rho : G \to L(V; V)$ be a representation of G on a Banach space V and $\Pi^\rho : F \to M$ the associated vector bundle. Let $\{y_t : 0 \leqslant t < \zeta\}$ be a $\mathcal{B}$-diffusion for an equivariant diffusion operator $\mathcal{B}$ over a cohesive diffusion operator $\mathcal{A}$. Set $x_t = p(y_t)$ and let $\Psi_t : F_{x_0} \to F_{x_t}, 0 \leqslant t < \zeta$ be the induced* *transformations on F. Then the local conditional expectation $\{\overline{\Psi}_t : 0 \leqslant t < \zeta\}$, for $\overline{\Psi}_t = \mathbf{E}\{\Psi_t | \sigma\{x_s : 0 \leqslant s < \zeta\}$ exists and is the solution of the covariant equation along $\{x_t : 0 \leqslant t < \zeta\}$:*

$$\frac{D}{\partial t}\overline{\Psi}_t = \Lambda^\rho \circ \overline{\Psi}_t$$

with Ψ_0 the identity map, $\Lambda^\rho : F \to F$ given by λ^ρ in Theorem 3.4.1 and where $\frac{D}{\partial t}$ refers to the semi-connection determined by $\mathcal{B}$.

Proof. From above and Proposition 4.10.1 we have

$$\Psi_t[(y_0, e)] = [(\tilde{x}_t \circ g_t^{\tilde{x}}, e)] = [(\tilde{x}_t, \rho(g_t^{\tilde{x}})^{-1}e]$$

and so $/\!/_t^{-1}\Psi_t[(y_0, e)] = [(y_0, \rho(g_t^{\tilde{x}})^{-1}e)]$. Now from the right invariance of $\mathcal{G}_t^\sigma$, for fixed path σ and time t, we can apply Baxendale's integrability theorem, [7], for the right action

$$
\begin{aligned}
G \times L(V; V) &\to L(V; V) \\
(g, T) &\mapsto \rho(g_t^\sigma)^{-1} \circ T
\end{aligned}
$$

to see $\mathbf{E}|\rho(g_t^\sigma)^{-1}|_{L(V;V)} < \infty$ for each σ, t and we have $\mathcal{E}(\sigma)_t \in L(V; V)$ given by

$$\mathcal{E}(\sigma)_t e = \mathbf{E}\rho(g_t^\sigma)^{-1}e.$$

By considering $/\!/_t^{-1}\overline{\Psi}_t$ we see that the local conditional expectation $\overline{\Psi}_t$ exists in $L(F_{x_0}; F_{x_t})$ and

$$\overline{\Psi}_t[(y_0, e)] = [(\tilde{x}_t, \mathcal{E}(x.)_t e)].$$

The computation in Theorem 3.4.1 shows that

$$\frac{d}{dt}/\!/_t^{-1}\overline{\Psi}_t[(y_0, e)] = \frac{d}{dt}[(y_0, \mathcal{E}(x.)_t e)] = [(y_0, \lambda^\rho(\tilde{x}_t)\mathcal{E}(x.)_t e)]$$

giving

$$\frac{D}{dt}\overline{\Psi}_t[(y_0, e)] = [(\tilde{x}_t, \lambda^\rho(\tilde{x}_t)\mathcal{E}(x.)_t e)] = \Lambda^\rho(x.)\overline{\Psi}_t[(y_0, e)]$$

as required. $\square$

Remark 4.11.2. Theorem 4.11.1 could also be used to identify the generator of the operator induced on sections of F^*, re-proving Theorem 3.4.1, since if $\phi \in \gamma F^*$, then $\mathbf{E}\phi \circ \Psi_t \chi_{t<\zeta} = \mathbf{E}\phi \circ \overline{\Psi}_t \chi_{t<\zeta}$ if the expectations exist, by Corollary 3.3.5 of [36]. The extra information in Theorem 4.11.1 is the existence of the conditional expectation. Baxendales' integrability theorem used for this applies in sufficient generality to give corresponding results for infinite dimensional G, for example in the situation arising in Chapter 8 below.

4.12 Notes

Noise-free observations

From the filtering viewpoint we could be considered to be dealing with *noise-free observations* as discussed by Joannides & LeGland in [54], though our assumption in this Chapter that the observations form a diffusion process make the situation much simpler.

Krylov-Veretennikov formula

The Krylov-Veretennikov formula, see Section 4.7, giving a Wiener chaos expansion for solutions of stochastic differential equations appeared in the 1976 article [103]. A version was described later by Elliott &Kohlmann in [29] in a form which is easily used for stochastic differential equations on manifolds, as described briefly in [32]. For an application to stochastic pde see [72]. It is the basis of the generalised stochastic flow theory of LeJan & Raimond, [63], [64]. In filtering theory it appeared in [77],[81], and other articles such as: [58], [74], and [73].

Skew-product decompositions, regular conditional probabilities and SDE

A basic measurability result concerning skew product decompositions obtained via by SDE's is given by John Taylor in [101].

Operators on differential forms

Suppose we have the stochastic differential equation $dx_t = X(x_t) \circ dB_t + A(x_t)\, dt$. on M with a solution flow $\xi_.$. The results of Theorem 4.11.1 combine with Corollary 3.4.9 to show that the operator on differential forms,

$$\mathcal{A}^\wedge := \frac{1}{2}\sum_{j=1}^{m}\mathbf{L}_{X^j}\mathbf{L}_{X^j} + \mathbf{L}_A,$$

and its semi-group $P_\cdot^\wedge$, described in the Notes on Chapter 2, Section 2.7, can be written as

$$\mathcal{A}^\wedge\phi \;=\; \frac{1}{2}\mathrm{trace}\,\check{\nabla}.\check{\nabla}.(\phi) - \frac{1}{2}\phi \circ \check{R}^k P_t^\wedge \phi(V) = \mathbf{E}\phi(W_t^k(V)) \qquad (4.30)$$

for ϕ a k-form and V an element of $\wedge^k T_{x_0}M$ some $x_0 \in M$. We are assuming the expectation exists. Here $\check{R}^k : \wedge^k TM \to \wedge^k TM$ is the generalised Weitzenböck curvature defined in Corollary 3.4.9 and $W_t^k : \wedge^k T_{x_0}M \to \wedge^k T_{\xi_t}M$ is the "damped" parallel translation given by

$$\frac{D}{dt}W_t^k(V) = -\check{R}^k\left(W_t^k(V)\right) \qquad (4.31)$$

using the covariant derivative of the connection adjoint to $\check{\nabla}$ in the sense of Section 3.3. This follows from the argument in Remark 4.11.2 since $W_t^k V = \mathbf{E}\{ \wedge^k T_{x_0}(V)|\xi_s(x_0), \quad 0 \leqslant s \leqslant t\}$ by Theorem 4.11.1 and Corollary 3.4.9. For a detailed discussion of this argument see [36], especially page 43, but beware of some incorrect signs there.

Chapter 5

Filtering with non-Markovian Observations

So far we have considered smooth maps $p : N \to M$ with a diffusion process $u_.$ on N mapping to a diffusion process $x_. = p(u_.)$ on M. From the point of view of filtering we have considered $u_.$ as the *signal* and $x_.$ as the *observation process*. However the standard set-up for filtering does not assume Markovianity of the observation process. Classically we have a signal $z_.$, a diffusion process on $\mathbf{R}^d$ or a more general space, and an observation process $x_.$ on some $\mathbf{R}^n$ given by an SDE of the form

$$dx_t = a(t, x_t, z_t)dt + b(t, x_t, z_t)dB_t \qquad (5.1)$$

where $B_.$ is a Brownian motion independent of the signal. To fit this into our discussion we will need to assume that the noise coefficient of the observation SDE does not depend on the signal other than through the observations, as well as the usual cohesiveness assumptions. We can take $N = \mathbf{R}^d \times \mathbf{R}^n$ and $M = \mathbf{R}^n$ with p the projection and $u_t = (z_t, x_t)$. To reduce to our Markovian case we can use the standard technique of applying the Girsanov-Maruyama theorem. Here we first carry this out in the general context of diffusions with basic symbols, as discussed in Section 2.4 and then show how it fits in with the classical situation. For simplicity we shall assume that the signal is a time-homogeneous diffusion, and that the coefficients in the observation SDE are also independent of time. The state spaces are taken to be smooth manifolds and the standard non-degeneracy assumptions on the observation process somewhat relaxed.

For other discussions about filtering with processes which have values in a manifold see Duncan,[26], Pontier &Szpirglas, [88], Davis &Spathopoulos, [24], Estrade,Pontier,& Florchinger, [41], and Gyöngy, [50].

K.D. Elworthy et al., *The Geometry of Filtering*, Frontiers in Mathematics,
DOI 10.1007/978-3-0346-0176-4_5, © Springer Basel AG 2010

5.1   Signals with Projectible Symbol

Using the notation and terminology of Section 2.4, suppose that our diffusion operator $\mathcal{B}$ on N is conservative and descends cohesively over $p : N \to M$ so that for a horizontal vector field b^H on N the diffusion operator $\tilde{\mathcal{B}} := \mathcal{B} - b^H$ lies over some cohesive $\mathcal{A}$. Choose such an $\mathcal{A}$ so that $\tilde{\mathcal{B}}$, and so $\mathcal{A}$, is also conservative: *we assume that this is possible.* Also choose a locally bounded one-form $b^\#$ on N with $2\sigma^{\mathcal{B}}(b^\#) = b^H$. This is possible since b^H is horizontal, and we can, and will, choose $b^\#$ to vanish on vertical tangent vectors and satisfy

$$b_y^\#(b^H(y)) = 2\sigma_y^{\mathcal{B}}(b_y^\#, b_y^\#) = \langle b^H(y), b^H(y) \rangle_y \qquad y \in N \qquad (5.2)$$

where $\langle -, - \rangle_y$ refers to the Riemannian metric on the horizontal tangent space induced by $2\sigma^{\mathcal{A}^H}$. This can be achieved by first choosing some smooth $\tilde{b} : N \to T^*M$ such that, in the notation of equation (2.28), $\sigma_{p(y)}^{\mathcal{A}}(\tilde{b}(y)) = b(y)$ for $y \in N$; and then taking $b^\#$ to be the pull-back of $\tilde{b}$ by p:

$$b_y^\#(v) = \tilde{b}(y)(T_y p(v)) \qquad y \in N.$$

Now set
$$Z_t = \exp\{-M_t^\alpha - \frac{1}{2}\langle M^\alpha \rangle_t\}$$

for $\alpha_t(u.) = b_{u_t}^\#$ where $u \in C([0, T]; N)$, our canonical probability space furnished with measures $\mathbf{P} := \mathbf{P}^{\mathcal{B}}$ and $\tilde{\mathbf{P}} := \mathbf{P}^{\tilde{\mathcal{B}}}$ and corresponding expectation operators $\mathbf{E}$ and $\tilde{\mathbf{E}}$.

Here and below we are using the notation of Proposition 4.1.1 with M^α etc referring to taking martingale parts with respect to $\mathbf{P}$ while $\tilde{M}^\alpha$ and $\int_0^t \alpha_s d\{y_s\}$ are with respect to $\tilde{\mathbf{P}}$. From the Girsanov-Maruyana-Cameron-Martin theorem (see the Appendix, Section 9.1), we know that $Z.$ is a martingale under $\mathbf{P}$ and the two measures are equivalent with

$$\frac{d\mathbf{P}_{y_0}^{\tilde{\mathcal{B}}}}{d\mathbf{P}_{y_0}^{\mathcal{B}}} = Z_T.$$

Suppose $f : N \to \mathbf{R}$ is bounded and measurable. We wish to find $\pi_t(f) : N \to \mathbf{R}, 0 \leqslant t \leqslant T$ where

$$\pi_t(f)(y_0) = \mathbf{E}_{y_0}\{f(u_t)|p(u_s), 0 \leqslant s \leqslant t\}.$$

Following the approach due to Zakai, consider the *unnormalised filtering process* $\hat{\pi}_t(f) : N \to \mathbf{R}$ given by

$$\hat{\pi}_t(f)(u_0) = \tilde{\mathbf{E}}_{u_0}\{f(u_t)Z_t^{-1} \mid p(u_s), 0 \leqslant s \leqslant t\}.$$

For completeness we state and prove the Kallianpur-Striebel formula, a version of Bayes' formula:

Lemma 5.1.1.
$$\pi_t(f)(u_0) = \frac{\hat{\pi}_t(f)(u_0)}{\hat{\pi}_t(1)(u_0)} \qquad \mathbf{P}_{u_0} - as.$$

Proof. Set $x_0 = p(u_0)$. Let $g : \mathcal{C}_{u_0}([0,T];N) \to \mathbf{R}$ be $\mathcal{F}_t^{x_0}$-measurable. Then

$$\mathbf{E}_{u_0}\{f(u_t)g(u_.)\} = \tilde{\mathbf{E}}\{\frac{1}{Z_t}f(u_t)g(u_.)\}$$

$$= \tilde{\mathbf{E}}\{\tilde{\mathbf{E}}\{\frac{1}{Z_t}f(u_t)|\mathcal{F}_t^{u_0}\}g(u_.)\}$$

$$= \mathbf{E}\{Z_t\tilde{\mathbf{E}}\{\frac{1}{Z_t}f(u_t)|\mathcal{F}_t^{u_0}\}g(u_.)\}. \tag{5.3}$$

Thus

$$\pi_t(f)(u_0) = \mathbf{E}\{Z_t|\mathcal{F}_t^{u_0}\}\hat{\pi}_t(f)(u_0).$$

Taking f constant shows that $\mathbf{E}\{Z_t|\mathcal{F}_t^{u_0}\}\hat{\pi}_t(1)(u_0) = 1$ and the result follows. $\square$

We can now go on to obtain the analogue of the Duncan-Mortensen-Zakai (DMZ) equation for the unnormalized filtering process, using the results of Section 4.8 on conditional laws:

Theorem 5.1.2. *For any C^2 function $f : N \to \mathbf{R}$, under $\tilde{\mathbf{P}}$,*

$$\hat{\pi}_t f(u_0) = f(u_0) + \int_0^t \hat{\pi}_s(\mathcal{B}f)(u_0)\, ds + \int_0^t \hat{\pi}_s\big(fb^{\#}(-)h_-\big)(u_0)d\{x_s\}$$

$$+ \int_0^t \hat{\pi}_s\big(df_-h_-\big)(u_0)d\{x_s\}; \tag{5.4}$$

$$\hat{\pi}_t f(u_0) = f(u_0) + \int_0^t \hat{\pi}_s(\mathcal{B}f)(u_0)\, ds + \int_0^t \langle \hat{\pi}_s(fb)(u_0), d\{x_s\}\rangle_{x_s}$$

$$+ \int_0^t \hat{\pi}_s\big(df_-h_-\big)(u_0)d\{x_s\} \tag{5.5}$$

where $x_s = p(u_s)$, $0 \leqslant s \leqslant \infty$ is the projection to M of the canonical process from u_0 on N, and h the horizontal lift map for the induced semi-connection.

Using an alternative notation:

$$\hat{\pi}_t f = \hat{\pi}_0 f + M_t^{\pi\left(fb^{\#}\circ h_{u.}\right),\mathcal{A}} + M_t^{\hat{\pi}.\left(df\circ h_{u.}\right),\mathcal{A}} + \int_0^t \hat{\pi}_s(\mathcal{B}f)ds. \tag{5.6}$$

Proof. Since we are working with $\tilde{\mathbf{P}}$ we will write $M^{b^{\#}}$ for $M^{b^{\#},\tilde{\mathcal{B}}}$, etc. Also Z_t^{-1} satisfies:

$$dZ_t^{-1} = Z_t^{-1}dM_t^{b^{\#}}$$

while

$$df(u_t) = dM_t^{df} + \tilde{\mathcal{B}}(f)(u_t)dt$$

giving

$$d\big(Z_t^{-1}f(u_t)\big) = Z_t^{-1}dM_t^{df} + Z_t^{-1}\tilde{\mathcal{B}}(f)(u_t)dt$$
$$+ f(u_t)Z_t^{-1}dM_t^{b^\#} + Z_t^{-1} + df_{u_t}(b^H(u_t))dt$$

since $dM_t^{df}dM_t^{b^\#} = \sigma^{\tilde{\mathcal{B}}}\big(df_{u_t}, b^\#\big)dt = df_{u_.}(b^H(u_t))dt.$ Thus

$$d\big(Z_t^{-1}f(u_t)\big) = Z_t^{-1}dM_t^{df} + Z_t^{-1}\mathcal{B}(f)(u_t)dt + f(u_t)Z_t^{-1}dM_t^{b^\#} + Z_t^{-1}.$$

We can now take conditional expectations using Proposition 4.3.5 since $\mathcal{B} - \mathbf{L}_{b^H}$ is over the cohesive operator $\mathcal{A}$ to complete the proof. $\square$

Lemma 5.1.3. *There are the following formulae for angle brackets:*

$$d\langle\hat{\pi}(1)\rangle_t = \langle\hat{\pi}_t(b), \hat{\pi}_t(b)\rangle_{x_t}^E dt, \tag{5.7}$$

$$d\langle\hat{\pi}(1), \hat{\pi}(f)\rangle_t = \langle\hat{\pi}_t(fb), \hat{\pi}_t(b)\rangle_{x_t}^E dt + \hat{\pi}_t(df \circ h_{u_.}) \circ \hat{\pi}_t(b(u_.))dt. \tag{5.8}$$

Proof. From the previous theorem

$$\langle\hat{\pi}(1), \hat{\pi}(f)\rangle dt = \big(dM_t^{\hat{\pi}(fb^\# \circ h),\mathcal{A}} + dM_t^{\hat{\pi}.(df \circ h_{u_.}),\mathcal{A}}\big)dM_t^{\pi(b^\# \circ h),\mathcal{A}}$$
$$= 2\sigma^{\mathcal{A}}\big(\hat{\pi}_t(fb^\# \circ h), \hat{\pi}_t(b^\# \circ h)\big)dt + 2\sigma^{\mathcal{A}}\big(\hat{\pi}_t(df \circ h_{u_.}), \hat{\pi}_t(b^\# \circ h)\big)dt$$
$$= \langle\hat{\pi}_t(fb), \hat{\pi}_t(b)\rangle_{x_t}dt + \hat{\pi}_t(df \circ h_{u_.}) \circ \hat{\pi}_t(b(u_.))dt,$$

since for any one-form ϕ on M we have:

$$\sigma^{\mathcal{A}}\big(\phi, \hat{\pi}_t(b^\# \circ h)\big) = \hat{\pi}_t\big(\langle\phi|_E, b^\# \circ h\rangle^{E^*}\big)$$
$$= \frac{1}{2}\hat{\pi}_t(\phi(b))$$
$$= \frac{1}{2}\phi(\hat{\pi}_t(b)).$$

This gives the second formula, from which comes the first. $\square$

We can now give a version of Kushner's formula in our context:

Theorem 5.1.4. *In terms of the probability measure* $\tilde{\mathbf{P}}$

$$\pi_t f = \pi_0 f + \int_0^t \pi_s \mathcal{B}(f)ds + \int_0^t \pi_s\big(df \circ h_{u_.}\big)\left[d\{x_s\} - \pi_s(b)ds\right] \tag{5.9}$$

$$+ \int_0^t \langle\pi_s(fb) - \pi_s(f)\pi_s(b), d\{x_s\} - \pi_s(b)\rangle_{x_s}ds.$$

Proof. From the definition and then Itô's formula:

$$d\pi_t(f) = d\left(\frac{\hat{\pi}_t(f)}{\hat{\pi}_t(1)}\right)$$
$$= \frac{d\hat{\pi}_t(f)}{\hat{\pi}_t(1)} - \frac{\hat{\pi}_t(f)d\hat{\pi}_t(1)}{(\hat{\pi}_t(1))^2} - \frac{d\hat{\pi}_t(f)d\hat{\pi}_t(1)}{(\hat{\pi}_t(1))^2}$$
$$+ \frac{\hat{\pi}_t(f)d\hat{\pi}_t(1)d\hat{\pi}_t(1)}{(\hat{\pi}_t(1))^3}.$$

Now substitute in the second formula of Theorem 5.1.2 and use the previous lemma. □

Note that $\hat{\pi}_t(f)$, b, and $\tilde{\mathbf{P}}$, depend on the choice of $\mathcal{A}$. We would like to have a version of formula (5.9) which is independent of such choices. First note that if $\mathcal{B} - b_1^H$ is over $\mathcal{A}_1$, and $\mathcal{B} - b_2^H$ is over $\mathcal{A}_2$, then the difference of the two vector fields on N descends to a vector field on M: if $g : M \to \mathbf{R}$ is smooth and $\tilde{g} = g \circ p : N \to \mathbf{R}$, then

$$(b_2^H - b_1^H)\tilde{g} = (\mathcal{B} - b_1^H)\tilde{f} - (\mathcal{B} - b_1^H)\tilde{g} = (\mathcal{A}_1 - \mathcal{A}_2)g.$$

Therefore if we set $b_0(z) = T_y p(b_2^H(y) - b_1^H(y))$ for $p(y) = z$, $z \in M$, then $\mathcal{A}_1 = \mathcal{A}_2 + \mathbf{L}_{b_0}$, and by Remark 4.1.4,

$$d\{x_s\}^{\mathcal{A}_2} = d\{x_s\}^{\mathcal{A}_1} + b_0 ds, \tag{5.10}$$

From this we see immediately that the symbols $d\{x_s\} - \pi_s(b)ds$, and $\pi_s(fb) - \pi_s(f)\pi_s(b)$ in formula (5.9) are in fact independent of the choice we made of $\mathcal{A}$. To relate to now classical concepts we next discuss the first of these in more detail.

5.2 Innovations and innovations processes

Keeping the notation above, for $\alpha \in L^2_{\mathcal{A}}$, so $\alpha_t \in T^*_{x_t} M$ for $0 \leqslant t < \infty$, define a real-valued process $I_t^\alpha : 0 \leqslant t < \infty$, the α-*innovations process* by

$$I_t^\alpha = \int_0^t \alpha_s \left(d\{x_s\}^{\mathcal{A}} - \pi_s b(u_.)ds\right) \tag{5.11}$$

A generalisation of a standard result about innovations processes is:

Proposition 5.2.1. *The process I^α is independent of the choice of $\mathcal{A}$. Under $\mathbf{P}^{\mathcal{B},u_0}$ it is an $\mathcal{F}_*^{x_0}$ martingale.*

Proof. The observations just made show that it is independent of the choice of $\mathcal{A}$. It is clearly also adapted to $\mathcal{F}_*^{x_0}$. To prove the martingale property note first that by Proposition 4.3.4 and formula (5.10)

$$\int_0^t \alpha_s d\{x_s\}^{\mathcal{A}} = \int_0^t p^*(\alpha_s)d\{u_s\}^{\mathcal{B}-\mathbf{L}_b H}$$
$$= \int_0^t p^*(\alpha_s)d\{u_s\}^{\mathcal{B}} - \int_0^t p^*(\alpha_s)b^H(u_s)ds$$
$$= \int_0^t p^*(\alpha_s)d\{u_s\}^{\mathcal{B}} - \int_0^t \alpha_s(b(u_s))ds.$$

From this we see that if $0 < r < t$ and $Z \in \sigma\{x_s : 0 \leqslant s \leqslant r\}$, then

$$\mathbf{E}^{\mathcal{B}} \chi_Z \left\{ \int_r^t \alpha_s \left(d\{x_s\}^{\mathcal{A}} - \pi_s b(u.) \right) ds \right\}$$
$$= \mathbf{E}^{\mathcal{B}} \chi_Z \left\{ \int_r^t \alpha_s \left(b(u_s) - \pi_s b(u.) \right) ds \right\} = 0$$

giving the required result.   □

If we fix a metric connection, Γ, on E, as described in Example 4.1.6 we can take the canonical Brownian motion, $B^{\Gamma, \mathcal{A}}$ say, on E_{x_0} determined by $\mathcal{A}$ and Γ. Then, by equation (4.8), we can write $d\{x_s\}^{\mathcal{A}} - \pi_s(b(u.))ds = /\!\!/_s dB^{\Gamma' \mathcal{A}} - \pi_s(b(u.))ds$. In terms of the the $\mathbf{P}$ Brownian motion, B^Γ, on E_{x_0}, which is the martingale part under $\mathbf{P}$ of the Γ- stochastic anti-development of x. we can define an E_{x_0}-valued process, $z_t^\Gamma : 0 \leqslant t < \infty$, by

$$z_t^\Gamma = B_t^\Gamma + \int_0^t (/\!\!/_s)^{-1}(b(u_s) - \pi_s(b(u.)))ds. \tag{5.12}$$

A candidate for the *innovations process* of our signal-observation system is the stochastic development , ν^Γ say, of z^Γ under Γ. This can be defined by using the canonical SDE on the orthonormal frame bundle of E, namely

$$d\tilde{\nu}_t = X(\tilde{\nu}_t)(\tilde{\nu}_0)^{-1} \circ dz_t$$

for a fixed frame ν_0 for E_{x_0}. Here

$$X(\mu)(e) = h_\mu^\Gamma(\mu(e))$$

for $\mu : \mathbf{R}^p \to E_m$ a frame in at some point $m \in M$, and $e \in \mathbf{R}^p$, for p the fibre dimension of E. The process ν^Γ is then the projection of $\tilde{\nu}$. on M. For example see [31]. It will satisfy the Stratonovich equation

$$d\nu_t^\Gamma = /\!\!/_t \circ dz_t \tag{5.13}$$

where the parallel translation is now along the paths of ν^Γ. Let $\Theta : C_0(M) \to C_0(M)$ be the map given by $\Theta(\sigma)_t = \nu^\Gamma(\sigma)_t$, treating z^Γ as defined on $C_0(M)$.

Let $\mathcal{D} = \mathcal{D}^\Gamma : C_0(T_{x_0}M \to C_{x_0}M$ be the stochastic development using Γ with inverse $\mathcal{D}^{-1}$. We will continue to assume that there is no explosion so that these maps are well defined. For example,

$$z(x.) = \mathcal{D}^{-1}\Theta(x.).$$

We define a semi-martingale, on M to be a Γ-*martingale* if it is the stochastic development using Γ of a local martingale, see the Appendix, Section 9.3.

Theorem 5.2.2. *For each metric connection Γ on E the innovations process ν^Γ is a Γ-martingale. If Γ is chosen so that the $\mathcal{A}$-diffusion process is a Γ-martingale under $\mathbf{P}^\mathcal{A}$, then for $\alpha : [0, \tau) \times C_{x_0}M \to T^*M$ which is predictable and lives over $x.$, provided the integrals exist,*

$$I^\alpha \circ \Theta(x.) = (\Gamma) \int_0^{\cdot} \alpha(\nu^\Gamma(x.).) d\nu^\Gamma(x.)_s - \int_0^{\cdot} \alpha(x.)_s \bar{b}(x.)_s ds \qquad (5.14)$$

where $\bar{b}(-)_s : C_{x_0} \to TM$ is the conditional expectation,

$$\bar{b}_s = \mathbf{E}\{b(u_s)|p(u.) = x.\},$$

and has $\bar{b}(x.)_s \in T_{x_s}M$ almost surely for all s.

Proof. The fact that ν^Γ is a Γ-martingale is immediate from the definition and Proposition 5.2.1. To prove the claimed identity note that our extra assumption on Γ implies that $/\!/_s^{-1} d\{x_s\}^\mathcal{A} = d(\mathcal{D}^{-1}(x.))_s$. Therefore

$$I^\alpha(x.) = \int_0^{\cdot} \alpha_s(x.) /\!/_s d\mathcal{D}^{-1}(x.)_s - \int_0^{\cdot} \alpha_s(x.) \bar{b}(x.)_s ds \qquad (5.15)$$

while by definition

$$(\Gamma) \int_0^{\cdot} \alpha_s d\nu^\Gamma_s(x.) = \int_0^{\cdot} \alpha_s(\nu^\Gamma_s(x.)) /\!/_s^{\nu^\Gamma(x.)} d(\mathcal{D}^{-1}(\nu^\Gamma(x.)))_s \qquad (5.16)$$

where the superscript on the parallel translation symbol indicates that it is along the paths $\nu^\Gamma(x.)$. Our identity follows. $\qquad \square$

Remark 5.2.3. (1) For Γ such that the $\mathcal{A}$-process is a Γ martingale we can easily see that Θ has an adapted inverse. Indeed its inverse is defined almost surely by

$$\Theta^{-1} = \mathcal{D} \circ \mathrm{Mart}^{\mathbf{P}^\mathcal{A}} \circ \mathcal{D}^{-1}$$

where $\mathrm{Mart}^{\mathbf{P}^\mathcal{A}}$ denotes the operation of taking the martingale part under the probability measure $\mathbf{P}^\mathcal{A}$.

(2) If we are given a connection Γ on E we could make our choice of $\mathcal{A}$ so that its diffusion process gives a Γ martingale. This specifies $\mathcal{A}$ uniquely and might be more natural sometimes, for example in the classical case with $M = \mathbf{R}^n$.

(3) The results and earlier discussion still hold if Γ is not a metric connection. However then $B^{\Gamma,\mathcal{A}}$ cannot be expected to be a Brownian motion. The connection could even be on TM rather than on E in which case $B^{\Gamma,\mathcal{A}}$ will be a local martingale in $T_{x_0}M$. This will be a natural procedure when $N = \mathbf{R}^n$, using the standard flat connection.

5.3   Classical Filtering

For an example of the situation treated above consider a signal process $(z_t, 0 \leqslant t \leqslant T)$ on $\mathbf{R}^d$ satisfying an SDE

$$dz_t = V(z_t, x_t)dW_t + \beta(z_t, x_t)dt \tag{5.17}$$

with $(x_t, 0 \leqslant t \leqslant T)$, the observation process, taking values in $\mathbf{R}^n$ and satisfying:

$$dx_t = X^{(1)}(x_t)dB_t + X^{(2)}(x_t)dW_t + b(z_t, x_t)dt. \tag{5.18}$$

Here B and W are independent Brownian motions of dimension q and p respectively. We then take $N = \mathbf{R}^d \times \mathbf{R}^n$ and $M = \mathbf{R}^n$, with $p : N \to M$ the projection. We set $u_t = (z_t, x_t)$ so that

$$
\begin{aligned}
\mathcal{B}f(z,x) = {} & \frac{1}{2}D^2_{1,1}f(V^i(z,x), V^i(z,x)) + D_1f(\beta(z,x)) \\
& + \frac{1}{2}D^2_{2,2}f(X^{(1),i}(x), X^{(1),i}(x)) + \frac{1}{2}D^2_{2,2}f(X^{(2),j}(x), X^{(2),j}(x)) \\
& + D_2f(z,x)(b(z,x)) + D^2_{1,2}f(z,x)(V^i(z,x), X^{(1),i}(z,x)) \tag{5.19}
\end{aligned}
$$

using the repeated summation convention where i goes from 1 to p and j from 1 to q, with the V^j referring to the components of V and similarly for $X^{(1),i}$ and $X^{(2),j}$. Also $D^2_{l,m}$ refers to the second partial Frechet derivative, mixed if $l \neq m$, etc.

The filtering problem would be to find $\mathbf{E}\{g(z_t) \mid x_s : 0 \leqslant s \leqslant t\}$ for suitable $g : \mathbf{R}^d \to \mathbf{R}$. This would fit in with the discussion above by defining $f : \mathbf{R}^d \times \mathbf{R}^n \to \mathbf{R}$ by $f(z,x) = g(z)$. Note that we have allowed feedback from the signal to the observation; usually only the special case where V and β are independent of x is considered. Also we have allowed the noise driving the signal to also affect the observations ("correlated noise"). This can give a non-trivial connection, in which case the terms involving horizontal derivatives of f will not vanish even for f independent of x. This vanishing would occur otherwise (i.e. for uncorrelated noise) so that in that case the formula in Theorem 5.1.2 reduces to the usual DMZ equation, for example as in [84] or [85]. For an up-to-date account of filtering theory see [2].

For an approach to dealing with the situation when the noise in the observation process has coefficients depending on the signal see [21].

Our basic assumptions are smoothness of the coefficients, non-explosion (for simplicity of exposition), and the cohesiveness of our observation process. By the latter we mean that for all $x \in \mathbf{R}^n$ and $z \in \mathbf{R}^d$ the image of the map $(e^1, e^2) \mapsto X^1(x)(e^1) + X^2(X)(e^2)$ from $\mathbf{R}^q \times \mathbf{R}^p$ to $\mathbf{R}^n$ contains $b(z, x)$ and has dimension independent of x. Some bounds are needed on b to ensure the existence of its conditional expectations.

To carry out the procedure for the signal and observation given above we must first identify the horizontal lift operator determined by $\mathcal{B}$. For this for each $x \in M$ let $Y_x : \mathbf{R}^n \to \mathbf{R}^{p+q}$ be the inverse of the restriction of the map $(e^1, e^2) \mapsto X^1(x)(e^1) + X^2(X)(e^2)$, from $\mathbf{R}^q \times \mathbf{R}^p$ to $\mathbf{R}^n$, to the orthogonal complement of its kernel. Then from Lemma 2.3.1 we see that the horizontal lift $h_u : \mathbf{R}^n \to \mathbf{R}^d \times \mathbf{R}^n$ is given by

$$h_u(v) = (V(z, x) \circ Y_x(v), v) \quad u = (z, x) \in \mathbf{R}^d \times \mathbf{R}^n. \qquad (5.20)$$

A natural choice of $\mathcal{A}$ is

$$\mathcal{A}(f)(x) = \frac{1}{2} D_{2,2}^2 f\left(X^{(1),i}(x), X^{(1),i}(x) \right) + \frac{1}{2} D_{2,2}^2 f\left(X^{(2),j}(x), X^{(2),j}(x) \right).$$

Having done that, the 'b' of our general discussion is just the drift $b : \mathbf{R}^d \times \mathbf{R}^n \to \mathbf{R}^n$ of our observation's stochastic differential equation. Moreover for suitable T^*M-valued processes α we have the α-innovations process

$$I_t^\alpha = \int_0^t \alpha(x_s) \left(X^{(1)}(x_s) dB_s + X^{(2)}(x_s) dW_s \right) + \int_0^t \alpha(x_s) \left(b(z_s, x_s) - \bar{b}(x)_s \right) ds,$$

where $\bar{b}(\sigma) = \mathbf{E}^{\mathcal{B}} \{ b(z_s, x_s) \mid x_r = \sigma_r \quad 0 \leqslant r \leqslant s \}$.

From Theorem 5.1.4, Kushner's formula, given smooth $g : \mathbf{R}^d \to \mathbf{R}$, one has

$$\pi_t g = g(z_0) + \int_0^t \left(\pi_s \frac{1}{2} D_{1,1}^2 g(-)(V^i(-), V^i(-)) + \pi_s D_1 g(-)(\beta(-)) \right) ds$$
$$+ I_t^{\pi_s \left(dg(-)V(-) \circ Y \right)} + I_t^{\langle \bar{g} b_s - \bar{g}_s \bar{b}_s, - \rangle_{x_s}}.$$

This can be compared, for example, with the formula given in the remark on page 85 of [85], following the proof of Proposition 2.2.5 there. Alternatively see [84].

Using the standard flat connection of R^n we get the innovations process given by

$$\nu_t = x_0 + \int_0^t \left(X^{(1)}(x_s) dB_s + X^{(2)}(x_s) dW_s \right) + \int_0^t \left(b(z_s, x_s) - \bar{b}(x)_s \right) ds.$$

5.4 Example: Another SPDE

Consider the stochastic partial differential equation on $L^2([0, 1]; \mathbf{R}^p)$:

$$du_t(x) = \Delta u_t(x) + \sum_{i=1}^m \Phi_i(x, u_t(x)) dB_t^i$$

where (B_t^i) are independent Brownian motions. For $p > 1$ it can be considered as a system of equations. One natural question is to find the law of u_t given that of $u_s(x_0), 0 \leqslant s \leqslant t$ for some given point x_0, or to find the conditional law of u_t given $u_s(x_0), 0 \leqslant s \leqslant t$. Here we indicate briefly how the approach we have been following may sometimes be applied to this or similar problems. A discussion of a similar but more straightforward problem was given in Section 4.9. For simplicity we take $p = 1$, so our "observations" process is one-dimensional: $M = \mathbf{R}$. As in Section 4.9 the state space N is taken to be a Sobolev space of sufficiently high regularity, $N = H^s([0.1]; \mathbf{R})$ for large s. The projection p is just evaluation at x_0.

Let $y_t = u_t(x_0)$. It satisfies:

$$dy_t = (\Delta u_t)(x_0)dt + \sum_{i=1}^{m} \Phi_i(x_0, y_t)dB_t^i.$$

Because of the drift term we cannot expect this to be Markovian so we will have to remove the term $(\Delta u_t)(x_0)dt$ by a Girsanov transformation.

Let (e_i) be the standard orthonormal base of $\mathbf{R}^m$. Define

$$\Phi : L^2([0,1]; \mathbf{R}) \times \mathbf{R}^m \to L^2([0,1]; \mathbf{R})$$

and

$$\tilde{\Phi} : \mathbf{R} \to \mathcal{L}(\mathbf{R}^m; \mathbf{R})$$

by

$$\Phi(u)(e)(x) = \sum_{i=1}^{m} \Phi_i(x, u(x))\langle e, e_i \rangle$$

and

$$\tilde{\Phi}(z)(e) := \sum_{i=1}^{m} \Phi_i(x_0, z)\langle e, e_i \rangle,$$

respectively. Consider $T_z\mathbf{R}$, identified with $\mathbf{R}$ and furnished with the metric induced by $\tilde{\Phi}(z)$:

$$\langle v_1, v_2 \rangle_z = \frac{v_1 v_2}{\sum_{i=1}^{m}(\tilde{\Phi}_i(z))^2}.$$

To have cohesivity and to be able to apply the Girsanov-Maruyama-Cameron-Martin theorem this must be well defined, i.e. the denominator must never vanish, and it must determine a non-explosive Brownian motion. If these conditions hold, we still have to be sure that the Girsanov transformed SPDE has solutions existing for all time and that we can apply the martingale method approach used in the proof of Theorem 9.1.4. Alternatively we can try to apply one of the standard tests to show that the local martingale which arises is a true martingale. First we apply Lemma 2.3.1 to obtain the horizontal lift map. For this we need the dual map $\tilde{\Phi}^*(z) : \mathbf{R} \to \mathbf{R}^m$ is given by:

$$\tilde{\Phi}^*(z)(1) = \frac{1}{\sum_{i=1}^{m}(\Phi_i(x_0, z))^2} \sum_{j=1}^{m} \Phi_j(x_0, z)e_j.$$

Then from equation (2.22) the horizontal lift $h_u : T_{u(x_0)}\mathbf{R} \to L^2([0,1];\mathbf{R})$ at a function u is given by

$$h_u(1)(x) = \Phi(x, u(x)) \circ \tilde{\Phi}^*(u(x_0)).$$

In particular a natural choice of drift b^H to be removed by the Girsanov-Maruyama-Cameron-Martin theorem, namely $b^H(u) = h_u(\triangle u(x_0))$, is given by

$$b^H(u)(x) = \frac{\sum_{j=1}^m \Phi_j(x_0, u(x_0))\Phi_j(x, u(x))}{\sum_{k=1}^n (\Phi_k(x_0, u(x_0)))^2} \triangle u(x_0). \tag{5.21}$$

Making the change of probability to $\tilde{\mathbf{P}}$ we see that our SPDE becomes

$$du_t(x) = \triangle u_t(x) - \frac{\sum_{j=1}^m \Phi_j(x_0, u_t(x_0))\Phi_j(x, u_t(x))}{\sum_{k=1}^n (\Phi_k(x_0, u_t(x_0)))^2} \triangle u_t(x_0)$$
$$+ \sum_{i=1}^m \Phi_i(x, u_t(x)) d\tilde{B}_t^i$$

for new, independent Brownian motions $\tilde{B}^1, \ldots, \tilde{B}^m$ and has the decomposition

$$du_t(x) = [\frac{\sum_{j=1}^m \Phi_j(x_0, u_t(x_0))\Phi_j(x, u_t(x))}{\sum_{k=1}^n (\Phi_k(x_0, u_t(x_0)))^2} \sum_{i=1}^m \Phi_i(x_0, u_t(x_0)) \, d\tilde{B}_t^i]$$
$$+ [\left(\triangle u_t(x) - \frac{\sum_{j=1}^m \Phi_j(x_0, u_t(x_0))\Phi_j(x, u_t(x))}{\sum_{k=1}^n (\Phi_k(x_0, u_t(x_0)))^2} \triangle u_t(x_0) \right) dt$$
$$+ \sum_{i=1}^m \left(\Phi_i(x, u_t(x)) - \frac{\sum_{j=1}^m \Phi_j(x_0, u_t(x_0))\Phi_j(x, u_t(x))}{\sum_{k=1}^n (\Phi_k(x_0, u_t(x_0)))^2} \Phi_i(x_0, u_t(x_0)) \right) d\tilde{B}_t^i].$$

In this decomposition the term in the first square brackets relates to the horizontal lift of the $\mathcal{A}$-process , while that in the second is the vertical component. They are independent (under $\tilde{\mathbf{P}}$), given u at x_0.

We could continue by applying the Kallianpur-Striebel formula, Lemma 5.1.1 or go directly to our version of Kushner's formula, Theorem 5.1.4. In that formula the operator $\mathcal{B}$ will be the infinite dimensional diffusion operator on $L^2([0,1];\mathbf{R})$ which is the generator of the solution of our SPDE, so there are extra analytical problems. There are cases where this machinery is not needed and the situation is fairly straightforward. For example:

(1) $\Phi_i(z, u) = \phi_i(z)$, where the vector $\{\phi_1(z), \ldots, \phi_m(z)\}$ never vanishes for any z. In this case y_t is basically Gaussian.

(2) The noise is one dimensional. In this case we see that the vertical noise term in our decomposition vanishes, so $u_t(z)$ is determined by $\{x_s : 0 \leqslant s \leqslant t\}$. In particular if $\Phi(z, u) = u$ with one-dimensional noise W_t, the SPDE has

solution $u_t(z) = e^{W_t - \frac{t^2}{2}} P_t u_0(z)$ where P_t refers to the Dirichlet heat kernel on $[0,1]$. For our cohesiveness assumption to hold in this case we would need to take N to consist of functions in H^s which are strictly positive. For u_0 in this space we have $W_t = \frac{1}{2}t + \log(\frac{u_t(x_0)}{P_t u_0(x_0)})$ which can be substituted back into the formula for $u_t(z)$.

However, the example

$$du_t = \Delta u_t + u_t dW_t + \phi(x) dB_t$$

is considerably more complicated even though it has the explicit solution:

$$u_t = e^{W_t - \frac{t}{2}} P_t u_0 + \int_0^t e^{W_t - W_s - \frac{t-s}{2}} P_{t-s} \phi dB_s,$$

P_t denoting the Dirichlet heat kernel as before.

For this example let us see, as an exercise, what terms will arise in Kushner's formula, Theorem 5.1.4,

$$\pi_t f = \pi_0 f + \int_0^t \pi_s \mathcal{B}(f) ds + \int_0^t \pi_s(df \circ h_{u.})\, [d\{x_s\} - \pi_s(b(u.))_s ds]$$

$$+ \int_0^t \langle \pi_s(fb) - \pi_s(f)\pi_s(b), d\{x_s\} - \pi_s(b) \rangle_{x_s},$$

for the conditional expectation of $f : N \to \mathbf{R}$ when $f(u) = u(z)$, for z our chosen point in $[0,1]$.

Note first that $\mathcal{B}f(u) = \Delta u(z)$. As in Section 4.9 denote $u(z)$ by $\delta_z(u)$, (so that now $f = \delta_z$). Recall from Section 4.9 that $\mathcal{B}\delta_z(u) = \Delta u(z)$.and $\mathcal{B}\delta_z = \delta_z \circ \Delta$. Similarly, since $b(u) = \Delta u(x_0)$, we have $b = \delta_{x_0} \circ \Delta$. Moreover,

$$h_u(1) = \frac{\phi(x_0)\phi + u(x_0)u}{\phi(x_0)^2 + u(x_0)^2} \quad \text{and} \quad b^H(u) = \Delta u(x_0) h_u(1).$$

Recalling that the generator on the base is given by

$$Ag(y) = (\phi(x_0)^2 + y^2)\frac{\partial^2 g}{\partial y^2}(y)$$

we have:

$$d\{y_s\}^A = \phi(x_0)dB_s + y_s dW_s + \Delta u(x_0)ds.$$

Also $\langle 1,1 \rangle_y = \frac{1}{\phi(x_0)^2 + y^2}$ and $\pi_s(b) = \pi_s(\delta_{x_0} \circ \Delta)$. The innovation increment $d\{y_s\}^A - \pi_s(b)ds$ is therefore given by

$$d\{y_s\}^A - \pi_s(b)(u_0)ds = \phi(x_0)dB_s + y_s dW_s + \Delta u_s(x_0)ds - \Delta \overline{u}_s(x_0)ds$$

where $\overline{u}_t = \mathbb{E}(u_t | \sigma(y_s, 0 \le s \le t))$, since then $\pi_s(\delta_z)(u_0) = \overline{u}_s(z)$ and $\pi_s(\Delta)(u_0) = \Delta \overline{u}_s(x_0)$.

We have $\pi_s(\mathcal{B}\delta_z) = \pi_s(\delta_z \circ \Delta)$. Also $\pi_s(\delta_z b) - \pi_s(\delta_z)\pi_s(b)$, which is

$$\pi_s((\delta_z)(\delta_{x_0} \circ \Delta)) - \pi_s(\delta_z)\pi_s(\delta_{x_0} \circ \Delta),$$

when applied to u_0 can be written

$$(\pi_s(\delta_z b) - \pi_s(\delta_z)\pi_s(b))(u_0) = \overline{u(z)\Delta u_s(x_0)} - \overline{u_s}(z)\Delta\overline{u_s}(x_0).$$

Finally, $\pi_s(df \circ h_-)(u_0) = \overline{h_{u_s}(1)(z)} = \frac{\phi(x_0)\phi(z) + u_s(x_0)\overline{u}_s(z)}{\phi(x_0)^2 + u_s(x_0)^2}$.

5.5 Notes

Noise-free observations

In this chapter we continued to deal with "noise-free observations", [54], but with the simplifying assumption that the quadratic variation of the observations process depends only on the signal through the observations. The more general case is discussed in [54] and [21]. We also reduced the standard filtering set up to our "noise-free" situation.

Chapter 6

The Commutation Property

In certain cases the filtering is in a sense trivial: the process decomposes into the observable and an independent process. From the geometric point of view this means the commutation of the vertical operator $\mathcal{B}^V$ and the horizontal operator $\mathcal{A}^H$. See Theorem 6.2.9 below.

For p a Riemannian submersion (defined in Chapter 7 below) with totally geodesic fibres and $\mathcal{B}$ the Laplacian, Berard-Bergery & Bourguignon [9] show that $\mathcal{A}^H$ and $\mathcal{B}^V$ commute. Their proof is based on the result of R.Hermann [51]

Theorem 6.0.1 (R. Hermann). *A Riemannian submersion $p : N \to M$ has totally geodesic fibres iff the Laplace-Beltrami operator of N commutes with all Lie derivations by horizontal lifts of vector fields on M.*

From this, and the Hörmander form representation of $\mathcal{A}^H$, it follows immediately that $\mathcal{A}^H$ with $\mathcal{B}^V$ will commute in their situation. In this section we consider some extensions of this and their consequences.

First, for $p : N \to M$ with a diffusion operator $\mathcal{B}$ over a cohesive $\mathcal{A}$, as usual, we will say that a vector field on N is *basic* if it is the horizontal lift of a section of E. From our Hörmander form representation of $\mathcal{A}^H$ we get the following extension of Berard-Bergery & Bourguignon's result:

Theorem 6.0.2. *For a diffusion operator $\mathcal{B}$ over a cohesive diffusion operator $\mathcal{A}$ the following are equivalent:*

(i) *$\mathcal{B}^V$ commutes with all Lie derivations by smooth basic vector fields of N;*

(ii) *the operators $\mathcal{B}$, $\mathcal{B}^V$, and $\mathcal{A}^H$ commute (on C^4 functions);*

(iii) *the operator $\mathcal{B}^V$ commutes with the horizontal lifts of the vector fields which appear in one Hörmander form representation of $\mathcal{A}$.*

Proof. It is clear that (i) implies (iii), and (iii) implies (ii). To show (ii) implies (i) observe that every section of E has the form $\sigma^{\mathcal{A}} \left(\sum_1^m \lambda^j df_j \right)$ since every one-form on M can be written as $\sum_1^m \lambda^j df_j$ for $\lambda^j : M \to \mathbf{R}$ and $f_j : M \to \mathbf{R}$ and some integer m. By definition of the connection this shows that every basic vector field on N has the form $\sum_1^m \lambda^j \circ p \, \sigma^{\mathcal{A}^H}(p^* df_j)$. It will therefore suffice to show that if (ii) holds, then $\mathcal{B}^V$ commutes with Lie differentiation by $\lambda \circ p \, \sigma^{\mathcal{A}^H}(p^* df)$ for all smooth $\lambda, f : M \to \mathbf{R}$.

For this assume (ii) holds and take a smooth $g : N \to \mathbf{R}$. By definition of the symbol and Remark 1.4.8:

$$
\begin{aligned}
2\mathcal{B}^V dg & \left(\lambda \circ p \, \sigma^{\mathcal{A}^H}(p^* df) \right) \\
&= 2\lambda \circ p \, \mathcal{B}^V dg \left(\sigma^{\mathcal{A}^H}(p^* df) \right) \\
&= \lambda \circ p \, \mathcal{B}^V \left(\mathcal{A}^H(f \circ p \, g) - f \circ p \, \mathcal{A}^H(g) - g\mathcal{A}^H(f \circ p) \right) \\
&= \lambda \circ p \, \left(\mathcal{A}^H(f \circ p)\mathcal{B}^V g - f \circ p \, \mathcal{A}^H \mathcal{B}^V g - (\mathcal{A}f) \circ p \, \mathcal{B}^V g \right) \\
&= 2\lambda \circ p \, d(\mathcal{B}^V g)\sigma^{\mathcal{A}^H}(p^* df) \\
&= 2d(\mathcal{B}^V g)\sigma^{\mathcal{A}^H}(\lambda \circ p \, p^* df)
\end{aligned}
$$

as required.   □

For the special case of an equivariant diffusion on a principal bundle as considered in Chapter 3 we can obtain a working criterion for commutativity: see also Example 6.2.13.

Corollary 6.0.3. *In the notation of Theorem 3.2.1, commutativity of $\mathcal{B}^V$ and $\mathcal{A}^H$ holds if and only if both α and β are constant along all horizontal curves. This holds if and only if $\mathcal{A}^H(\alpha^{i,j}) = 0$ and $\mathcal{A}^H(\beta^k) = 0$ for all i, j, k.*

Proof. First note that each vector field A_k^* commutes with all basic vector fields. Indeed if V is basic it is equivariant and so

$$(R_{\exp t A_k})_*(V) = V \qquad t > 0.$$

Differentiating in t at $t = 0$ gives the required commutativity. Thus the operators $\mathcal{L}_{A_k^*}$ are invariant under flows of basic vector fields and so for $\mathcal{B}^V$ to commute with basic vector fields the coefficients α and β must be constant along their flows. By the theorem this gives the first result since any horizontal curve can be considered as an integral curve of a (possible time dependent) basic vector field.

Clearly, from the Hörmander form of $\mathcal{A}^H$, if this holds both α and β are $\mathcal{A}^H$-harmonic. The converse holds since, from above, $\mathcal{A}^H$ commutes with all of the vertical vector fields $\mathcal{L}_{A_k}^*$.   □

The Corollary is applied to derivative flows in Example 6.2.13 of Section 6.2 below.

Hermann proved that a Riemannian submersion with totally geodesic fibres , see Section 9.4, has the natural structure of a fibre bundle with group the isometry group of a typical fibre:

Theorem 6.0.4 (R. Hermann). *If N is a complete Riemannian manifold and ϕ : $N \to M$ is a C^∞ Riemannian submersion, then ϕ is a locally trivial fibre space. If in addition the fibres of ϕ are totally geodesic submanifolds of N, ϕ is a fibre bundle with structure group the Lie group of isometries of the fibre.*

An analogous result given the hypothesis of Theorem 6.0.2 together with some completeness and hypoellipticity conditions is proved in Theorem 6.2.9 below.

Before that we consider when the associated semi-groups commute.

6.1 Commutativity of Diffusion Semigroups

It is well known that in general the commutativity of two diffusion generators (on C^4 functions) does not imply that of their associated semi-groups. One reference is [90] page 273 where an example they ascribe to Nelson is given. Here is a minor modification of that construction:

Cut $\mathbf{R}^2$ along the positive x-axis and remove the origin. Take a copy C, say, of $(0, \infty) \times [0, \infty)$ and glue it along the cut to the lower part of the cut plane, identifying $(0, \infty) \times \{0\}$ in C with the positive x-axis. This gives a version of the plane but with two copies of the positive quadrant, the original one now incomplete along the positive x-axis and the second one, C, incomplete along the positive y-axis, and with the origin missing. On this we have naturally defined vector fields X^1 given by $\frac{\partial}{\partial x}$ and X^2 given by $\frac{\partial}{\partial y}$. These certainly commute. However their associated semi-groups do not, as can be seen by starting at the point $(-1, -1)$ moving along the X_1-trajectory for time 2 and then along the X^2 trajectory for the same amount of time. We end up at the point $(1, 1)$ of copy C. However if we had changed the order of the vector fields we would be at $(1, 1)$ of the original positive quadrant.

A more geometrically satisfying construction would be, as Nelson, to use the double covering of the punctured plane as state space with similarly behaved vector fields.

Though it will not be used in the sequel here is an easy positive result:

Proposition 6.1.1. *Let A_1 and A_2 be diffusion operators with associated semi-groups $\{P_t^1\}_{t>0}$ and $\{P_t^2\}_{t>0}$ acting as strongly continuous semi-groups on a Banach space E of functions which contains the C^2 functions with compact support. Let $\mathcal{G}_1$ and $\mathcal{G}_2$ be the corresponding generators, (closed extensions of the restrictions of A_1 and A_2 to the space of C^2 functions with compact support). Assume there is a core $\mathcal{C}_2$ for $\mathcal{G}_2$ consisting of bounded C^∞ functions such that for $f \in \mathcal{C}_2$:*

(i) *For all $t > 0$ the function $P_t^1 f$ is C^4.*

(ii) *$A_2 \frac{1}{t}(P_t^1 f - f)$ is uniformly bounded in $t \in (0, 1)$ and in space, and it converges pointwise to $A_2 A_1 P_t^1 f$ as $t \to 0+$.*

(iii) *$A_2 P_t^1 f$ is uniformly bounded in $t \in (0, 1)$ and in space.*

Then commutativity of P_t^1 with P_s^2, $0 \leqslant s$, $0 \leqslant t$ follows from commutativity of A_1 with A_2 on C^4 functions. Moreover if this holds, the semi-group $\{P_t^{A_1+A_2}\}_{t>0}$ associated to $A_1 + A_2$ satisfies

$$P_t^{A_1+A_2} = P_t^1 P_t^2.$$

Proof. Let $f : M \to \mathbf{R}$ be in C_2. We show first that

$$A_2 P_t^1 f = P_t^1 A_2 f. \tag{6.1}$$

For this set $V_t = A_2 P_t^1 f$. Then, by hypothesis (ii),

$$\frac{\partial}{\partial t} V_t = A_2 A_1 P_t^1 f$$

$$= A_1 V_t \tag{6.2}$$

by commutativity. By assumption (ii) we know V_s is bounded uniformly in $s \in [0, t]$ for any $t > 0$. However there is a unique C^2 and uniformly bounded solution, $P^1 V_0$, to any diffusion equations such as (6.2) with given smooth bounded initial condition V_0 (as is easily seen by the standard use of Itô's formula applied to V_{t-s} acting on a diffusion process with generator A_1). This gives

$$A_2 P_t^1 f = P_t^1 V_0 = P_t^1 A_2 f$$

as required. Now suppose $f \in \mathrm{Dom}(\mathcal{G}_2)$. By assumption there is a sequence $\{f_n\}_n$ of functions in C_2 converging in $\mathcal{G}_2$-graph norm to f. Then $P_t^1 A_2 f_n \to P_t^1 \mathcal{G}_2 f$ and $P_t^1 f_n \to P_t^1 f$. Equation (6.1) therefore shows that $P_t^1 f \in \mathrm{Dom}(\mathcal{G}_2)$ and we have

$$\mathcal{G}_2 P_t^1 \supset P_t^1 \mathcal{G}_2. \tag{6.3}$$

Next, for $f \in \mathrm{Dom}(\mathcal{G}_2)$, and our fixed $t > 0$ set $W_s = P_t^1 P_s^2 f$. Since the convergence of $\frac{1}{\epsilon}\{P_{s+\epsilon}^2 f - P_s^2 f\}$ to $\mathcal{G}_2 P_s^2 f$ is in E we see, using equation (6.3),

$$\frac{\partial}{\partial s} W_s = P_t^1 \mathcal{G}_2 P_s^2 f = \mathcal{G}_2 P_t^1 P_s^2 f = \mathcal{G}_2 W_s$$

since $P_s^2 f \in \mathrm{Dom}(\mathcal{G}_2)$. In particular $W_s \in \mathrm{Dom}(\mathcal{G}_2)$.

Although now it is not clear that W is C^2 we see from this that $\frac{\partial}{\partial u} P_u^2 W_{s-u} = 0$ for $0 < u < s$, giving

$$P_t^1 P_s^2 f = P_0^2 W_s = P_s^2 W_0 = P_s^2 P_t^1 f$$

for $0 \leqslant s \leqslant t$. For $s > t$ it is now only necessary to use the semigroup property of P^2, to commute with P_t^1 portion by portion.

Finally since $P_t^2 f \in \mathrm{Dom}(\mathcal{G}_2)$ the above gives

$$\frac{\partial}{\partial t} P_t^1 P_t^2 f = A_1 P_t^1 P_t^2 f + P_t^1 A_2 P_t^2 f$$

$$= (A_1 + A_2) P_t^1 P_t^2 f$$

and we can repeat the second arguement showing uniqueness of solutions of the diffusion equation to obtain $P_t^{A_1+A_2} f = P_t^1 P_t^2 f$. $\square$

Remark 6.1.2. Condition (i) does not always hold. A simple example is when the state space is $\mathbf{R}^2 - \{(0,0)\}$ and the operator is $\frac{\partial^2}{\partial x^2}$. The standard positive result for degenerate operators on $\mathbf{R}^n$ is due to Oleĭnik, [82].

6.2 Consequences for the Horizontal Flow

For our standard set-up of $p : N \to M$ with diffusion operator $\mathcal{B}$ over a cohesive $\mathcal{A}$, let P^V and P^H denote the semi-groups , generated by the vertical and horizontal components of B, and let $p_t^V(u, -), t \leqslant 0, u \in N$, be the transition probabilities of P^V. If we set $N_x = p^{-1}(x)$ for $x \in M$, then $p_t^V(u, -)$ will be a probability measure on $N_{p(u)}^+$, the union of $N_{p(u)}$ with Δ.

For $\mathbf{P}_{x_0}^{\mathcal{A}}$-almost all $\sigma \in C_{x_0} M^+$ for each $x_0 \in M$ there are measurable maps, the *stochastic holonomy* or *stochastic parallel translation*

$$/\!/_t^\sigma : N_{x_0}^+ \to N_{\sigma_t}^+$$

such that for each $u \in N_{x_0}$ the process $(t, \sigma) \mapsto /\!/_t^\sigma(u)$ is an $\mathcal{A}^H$-diffusion and is over σ. These can be obtained, for example, by taking a stochastic differential equation, such as equation (4.22),

$$dx_t = X(x_t) \circ dB_t + A(x_t)dt$$

for our $\mathcal{A}$-diffusion. Let $Y_x : E_x \to \mathbf{R}^m$ be the adjoint (and right inverse) of $X(x)$, each $x \in M$. Then consider the SDE on N,

$$dy_t = \tilde{X}(y_t)Y(\sigma_t) \circ d\sigma_t$$

and let $(t, \sigma) \mapsto /\!/_t^\sigma$ be the restriction of its flow to N_{x_0}, augmented by mapping the coffin state, Δ, to itself. This SDE is canonical since by equation (2.22) it can be rewritten as

$$dy_t = h_{y_t} \circ d\sigma_t$$

for h the horizontal lift map of Proposition 2.1.2.

We will often need to assume that the lifetime of this diffusion is the same as that of its projection on M:

Definition 6.2.1. The semi-connection induced by $\mathcal{B}$ is said to be *stochastically complete* if

$$C_{u_0}^p M^+ := \{\sigma : [0, \infty) \to M^+ : \lim_{t \to \zeta} p(u_t) = \Delta \text{ when } \zeta(u) < \infty\}$$

has full $\mathbf{P}_{u_0}^{\mathcal{A}^H}$ measure for each $u_0 \in N$ or equivalently if the lifetimes satisfy

$$\zeta^M(\sigma) = \zeta^M(/\!/^\sigma u_0)$$

for $\mathbf{P}_{u_0}^{\mathcal{A}}$-almost all paths σ, and all $u_0 \in N_{\sigma(0)}$.

The semi-connection is said to be *strongly stochastically complete* if also we can choose a version of $/\!/_t^\sigma : N_{\sigma(0)} \to N_{\sigma(t)}$ which is a smooth diffeomorphism whenever $\sigma(0)$ is a regular value of p and $t < \zeta(\sigma)$.

Note that strong stochastic completeness of the connection will hold whenever the fibres of p are compact by the basic properties of the domains of local flows of SDE, [59], [30]. It also holds if the stochastic horizontal differential equation is strongly p-complete in the sense of Li [68] for $p = \dim(N) - \dim(M)$.

Proposition 6.2.2. *Suppose the semi-groups P^V and P^H commute and stochastic completeness of the connection holds. Then the horizontal flow preserves the vertical transition probabilities in the sense that for all positive s and $0 < t < \zeta(\sigma)$,*

$$(/\!/_t^\sigma)_* p_s^V (u_0, -) = p_s^V (/\!/_t^\sigma u_0, -) \tag{6.4}$$

for all $u_0 \in N_\sigma$ for $\mathbf{P}^{\mathcal{A}}$-almost all σ. Equivalently for any bounded measurable $h : N \to \mathbf{R}$ we have $\mathbf{P}^{\mathcal{A}}$-almost surely;

$$P_s^V (h \circ /\!/_t^\sigma) (u_0) = P_s^V h (/\!/_t^\sigma (u_0)). \tag{6.5}$$

Proof. It suffices to show that given any finite sequence $0 \leqslant t_1 \leqslant t_2 \leqslant \cdots \leqslant t_k < t$, bounded measurable $f_j : M \to \mathbf{R}$, $j = 1, \ldots, k$ and bounded measurable $h : N \to \mathbf{R}$, if $u_0 \in N_{x_0}$, then

$$\mathbf{E}_{x_0} \left\{ f_1(\sigma_{t_1}) \ldots f_k(\sigma_{t_k}) \chi_{t < \zeta(\sigma)} \cdot P_s^V (h \circ /\!/_t^\sigma) (u_0) \right\}$$
$$= \mathbf{E}_{x_0} \left\{ f_1(\sigma_{t_1}) \ldots f_k(\sigma_{t_k}) \chi_{t < \zeta(\sigma)} \cdot P_s^V (h) (/\!/_t^\sigma (u_0)) \right\} \tag{6.6}$$

where χ_Z denotes the indicator function of a set Z. To see this set $\tilde{f}_j = f_j \circ p : N \to \mathbf{R}$. Then the left-hand side of (6.6) is

$$\mathbf{E}_{x_0} \left\{ \tilde{f}_1(/\!/_{t_1}^\sigma(u_0)) \ldots \tilde{f}_k(/\!/_{t_k}^\sigma(u_0)) \cdot \chi_{t < \zeta(\sigma)} \cdot P_s^V (h \circ /\!/_t^\sigma)(u_0) \right\}$$
$$= \mathbf{E}_{x_0} \left\{ P_s^V \left(\tilde{f}_1(/\!/_{t_1}^\sigma(u_0)) \ldots \tilde{f}_k(/\!/_{t_k}^\sigma(u_0)) \cdot \chi_{t < \zeta(\sigma)} \cdot h \circ /\!/_t^\sigma \right) (u_0) \right\}$$
$$= P_s^V \left(P_{t_1}^H \tilde{f}_1 \ldots \left(P_{t_k - t_{k-1}}^H \tilde{f}_k \right) P_{t - t_k}^H h \right) (u_0)$$
$$= \left(P_{t_1}^H \tilde{f}_1 \ldots P_{t_k - t_{k-1}}^H \tilde{f}_k P_{t - t_k}^H P_s^V h \right) (u_0)$$

which reduces to the right-hand side of (6.6).   □

Remark 6.2.3. Assuming strong stochastic completeness of our semi-connection let $\{z_t : 0 \leqslant t < \zeta(p(u_.)\}$ be a semi-martingale in N with

$$p(z_t) = x_t := p(u_t) : 0 \leqslant t < \zeta(p(u_.)).$$

If x_0 is a regular value of p we have the Stratonovich equation:

$$d/\!/_t^{-1} z_t = T/\!/_t^{-1} \circ dz_t - T/\!/_t^{-1} (h_{z_t} \circ dx_t) \tag{6.7}$$

where $/\!\!/_t$ refers to $/\!\!/_t^x$. To see this, for example set $b_t = /\!\!/_t^{-1} z_t$ and observe that

$$dz_t = d(/\!\!/_t b_t) = T/\!\!/_t \circ db_t + h_{/\!\!/_t b_t} \circ dx_t.$$

Now assume that our induced semi-connection is strongly stochastically complete. For a regular value x_0 of p and $u_0 \in N_{x_0}$ define a process $\alpha^{u_0} : [0, \infty) \times \mathcal{C}_{u_0} N^+ \to N_{x_0}^+$ by

$$\alpha_t^{u_0}(u) = \alpha_t(u) = (/\!\!/_t^{p(u)})^{-1} u_t \tag{6.8}$$

if $u \in \mathcal{C}_{u_0} N$ with $t < \zeta(u)$ and define $\alpha_t^{u_0}(u) = \Delta$ if $t \geqslant \zeta(u)$. Note that α_t may not go out to infinity in N_{x_0} as t increases to its extinction time.

Also define

$$/\!\!/_s^*(\mathcal{B}^V)(f) = \mathcal{B}^V(f \circ /\!\!/_s) \circ /\!\!/_t^{-1}$$

to obtain a random time-dependent diffusion operator $/\!\!/_s^*(\mathcal{B}^V)$ on each fibre over a regular value of p.

Lemma 6.2.4. *In the notation of equation (4.24) we have the Itô equation for* $\alpha_t := \alpha_t^{u_0}$:

$$\nabla^V \qquad d\alpha_t = T/\!\!/_t^{-1} V(/\!\!/_t \alpha_t) dW_t + T/\!\!/_t^{-1} V^0(/\!\!/_t \alpha_t) dt$$

which we can write as:

$$\nabla^V \qquad d\alpha_t = /\!\!/_t^* V(\alpha_t) dW_t + /\!\!/_t^* V^0(\alpha_t) dt. \tag{6.9}$$

In particular for $f : N \to \mathbf{R}$ *in* C^2

$$M_t^{df,\alpha} := f(\alpha_t) - \int_0^t /\!\!/_s^*(\mathcal{B}^V)(f)(\alpha_s) ds \tag{6.10}$$

is a local martingale.

Proof. Formula (6.9) is immediate from equations (4.24) and (6.7). That $M^{df,\alpha}$ is a local martingale follows immediately using the properties of pull-backs under diffeomorphisms of Lie derivatives when V is C^1, and by going to local co-ordinates otherwise. □

Lemma 6.2.5.

Suppose $\mathcal{A}$ has Hörmander form

$$\mathcal{A} = \frac{1}{2} \sum_{j=1}^m \mathbf{L}_{X^j} \mathbf{L}_{X^j} + \mathbf{L}_X^0.$$

Let $/\!\!/^j$ be the flow of the horizontal lift $\tilde{X}^j$ of the vector field $\widetilde{X}^j$ for $j = 0, \ldots, m$ and represent $/\!\!/.$ by the flow of the SDE:

$$dz_s = \sum_{j=1}^m \widetilde{X}^j(z_s) \circ dB^j + \widetilde{X}^0(z_s).$$

Then $\frac{d}{ds}(/\!/_s^j)^*\mathcal{B}^V$ is the commutator $[\mathbf{L}_{X^j},(/\!/_s^j)^*\mathcal{B}^V]$. Consequently for any C^3 function $f : N \to \mathbf{R}$ and $u_0 \in N$ we have the Stratonovich formula:

$$d\left((/\!/_s)^*(\mathcal{B}^V)(f)(u_0)\right) = \sum_{j=1}^{m}[\mathbf{L}_{X^j},(/\!/_s^j)^*\mathcal{B}^V](f)(u_0) \circ dB^j$$

$$+ [\mathbf{L}_{X^0},(/\!/_s^0)^*\mathcal{B}^V](f)(u_0)\ dt. \qquad (6.11)$$

Proof. From a standard result for any smooth vector field V on N we have $\frac{d}{ds}(/\!/_s^i)^*\mathbf{L}_V = [X^j, V]$. We can take a representation for $\mathcal{B}^V$, as in equation (1.7) of Section 1.1,

$$\mathcal{B}^V = \frac{1}{2}\sum_{ij=1}^{N} a^{ij}(\cdot)\mathbf{L}_{V^i}\mathbf{L}_{V^j} + \mathbf{L}_{V^0}$$

with smooth a_{ij} and vector fields V^j. If the a_{ij} all vanish, so that we have a first-order operator, we can apply the standard result. Otherwise we can suppose $V^0 = 0$ and just consider $\mathcal{B}^V = a(\cdot)\mathbf{L}_{V^1}\mathbf{L}_{Y^2}$. Then

$$\frac{d}{ds}(/\!/_s^j)^*\mathcal{B}^V = \frac{d}{ds}\left(a(/\!/_s^j(\cdot))(/\!/_s^j)^*(\mathbf{L}_{V^1})(/\!/_s^j)^*(\mathbf{L}_{V^2})\right)$$

$$= \mathbf{L}_{X^j}(a)(/\!/_s^j(\cdot))(/\!/_s^j)^*(\mathbf{L}_{V^1})(/\!/_s^j)^*(\mathbf{L}_{V^2})$$

$$+ a(/\!/_s^j(\cdot))\left((/\!/_s^j)^*([\mathbf{L}_{X^j},\mathbf{L}_{V^1}])(/\!/_s^j)^*\mathbf{L}_{V^2} + \mathbf{L}_{V^1}(/\!/_s^j)^*[\mathbf{L}_{X^j},\mathbf{L}_{V^2}]\right).$$

Now $(/\!/_s^j)^*[\mathbf{L}_{X^j},\mathbf{L}_{V^j}] = [\mathbf{L}_{X^j},(/\!/_s^j)^*\mathbf{L}_{V^j}]$ since Lie brackets commute with diffeomorphisms and any vector field is invariant under its flow. Expanding the brackets yields

$$\frac{d}{ds}(/\!/_s^j)^*\mathcal{B}^V = \mathbf{L}_{X^j}(a)(/\!/_s^j(\cdot))(/\!/_s^j)^*(\mathbf{L}_{V^1})(/\!/_s^j)^*(\mathbf{L}_{V^2})$$

$$+ a(/\!/_s^j(\cdot))[\mathbf{L}_{X^j},(/\!/_s^j)^*(\mathbf{L}_{V^1}\mathbf{L}_{V^2})]$$

$$= [\mathbf{L}_{X^j},(/\!/_s^j)^*\mathcal{B}^V]$$

as required. $\square$

Remark 6.2.6. Since $(/\!/_s^j)^*\mathcal{B}^V$ is determined by the operators $\mathcal{A}$ and $\mathcal{B}$, at least for fibres over regular values of p, it is independent of the choice of Hörmander form for $\mathcal{A}$. That means that the stochastic differential $d\left((/\!/_s)^*(\mathcal{B}^V)(f)(u_0)\right)$ should also be expressible in terms of the operators using the approach of Section 4.1.

Definition 6.2.7. For a regular value x_0 of p. We say $\mathcal{B}^V$ is *stochastically holonomy invariant at* x_0 if on N_{x_0} we have $/\!/_t^*(\mathcal{B}^V) = \mathcal{B}^V$ for all $0 \leqslant t < \zeta^M$ with probability 1. If this holds for all regular values x_0, then we say $\mathcal{B}^V$ is *stochastically holonomy invariant*. Similarly we say B^V is *holonomy invariant at* x_0 if the corresponding result holds for parallel translation along any piecewise C^1 curve starting at x_0 in M, and is *holonomy invariant* if this holds for all regular values x_0.

Remark 6.2.8. 1. If the $\mathcal{A}$-diffusion on M is represented by a stochastic differential equation we can lift that equation to N and obtain a local flow $\eta_t^H : 0 \leqslant t < \zeta^H(-)$ where $\zeta^H(y) : y \in N$ gives its explosion times; so that with probability 1 η_t^H is defined and smooth on the open set $\{y \in N : t \leqslant \zeta^H(y)$, see [59] or [30]. We can say that $\mathcal{B}^V$ is invariant under the horizontal flow if for all C^2 functions $f : N \to \mathbf{R}$ we have

$$\mathcal{B}^V(f) \circ \eta_t = \mathcal{B}^V(f \circ \eta_t)$$

on $\{y \in N : t \leqslant \zeta^H(y)$, almost surely, for all $t > 0$. This does not require strong stochastic completeness of the semi-connection, nor do we have to restrict attention to fibres over regular values. On the other hand if it holds, and given such strong stochastic completeness, if x_0 is a regular value it follows that N_{x_0} lies in $\{y \in N : t \leqslant \zeta^H(y)$ for all $t < \zeta^M(x_0)$ and that we have stochastic holonomy invariance at x_0.

2. Assume completeness of the semi-connection. If $\mathcal{A}$ satisfies the standard Hörmander condition, or more generally if the space $\mathcal{D}^0(x_0)$, as in Section 2.6, is all of M, then holonomy invariance at x_0 implies holonomy invariance. This follows since concatenation of paths gives composition of the corresponding parallel translations and the conditions imply that any two points can be joined by a smooth path with derivatives in E. Moreover by Theorem 2.6.1 every point is a regular value and so given also strong stochastic completeness of the connection, from the theorem below we see that holonomy invariance of B^V at one point implies it is invariant under the horizontal flow induced by any SDE on M which gives one-point motions with generator $\mathcal{A}$. The same holds for stochastic holonomy invariance: see Theorem 6.2.9 below.

Theorem 6.2.9. *Suppose the induced semi-connection is complete and strongly stochastically complete, and x_0 is a regular value of p. Then the following are equivalent:*

$(\mathrm{i}x_0)$ *For all $u_0 \in N_{x_0}$ and for any $\mathcal{F}^\alpha$-stopping time τ with $\tau(\alpha(u)) < \zeta(p(u))$, the process $\{\alpha_t : 0 \leqslant t < \tau\}$ is independent of $\mathcal{F}^{x_0}$;*

$(\mathrm{ii}x_0)$ $\mathcal{B}^V$ *is stochastically holonomy invariant at x_0;*

$(\mathrm{iii}x_0)$ $\mathcal{B}^V$ *is holonomy invariant at x_0;*

$(\mathrm{iv}x_0)$ $\mathcal{B}^V$ *and $\mathcal{A}^H$ commute at all points of $\overline{\mathcal{D}^0(x_0)}$;*

$(\mathrm{v}x_0)$ P^V *and P^H commute at all points of $\overline{\mathcal{D}^0(x_0)}$.*

If the above hold at some regular value x_0 they hold for all elements in $\mathcal{D}^0(x_0)$. Moreover α^{u_0} is a Markov process on N_{x_0} with generator $\mathcal{B}^V$.

Proof. We will show that $(\mathrm{i}x_0)$ is equivalent to $(\mathrm{ii}x_0)$ which implies $(\mathrm{iv}x_0)$. Then $(\mathrm{iv}x_0)$ implies $(\mathrm{iii}y)$ for all $y \in \mathcal{D}^0(x_0)$ which implies $(\mathrm{v}x_0)$. Finally we show $(\mathrm{v}x_0)$ implies $(\mathrm{ii}y)$ for all $y \in \mathcal{D}^0(x_0)$.

Assume (ix_0) holds. Let $f : N_{x_0} \to \mathbf{R}$ be smooth with compact support. Then the local martingale $M^{df,\alpha}$ given by formula (6.10) is a martingale and from equation (6.9) we see that

$$\mathbf{E}\{M^{df,\alpha}|\mathcal{F}^{x_0}\} = f(u_0).$$

Therefore, for $\mathbf{P}^{x_0}$-almost all σ in $C_{x_0}M$,

$$\mathbf{E}\{f(\alpha_t)\} = \mathbf{E}\{f(\alpha_t)|p(u_.) = \sigma\} = f(u_0) + \int_0^t \mathbf{E}\{(/\!/_s^\sigma)^*(\mathcal{B}^V)(f)(\alpha_s)\}ds. \quad (6.12)$$

Also, in the notation of equation (6.9), with the obvious notation for the filtrations generated by our processes, we have $\mathcal{F}_t^{\alpha_.} \subset \mathcal{F}_t^{W_.} \wedge \mathcal{F}_t^{x_0}$ and $\mathcal{F}_t^{W_.} \subset \mathcal{F}_t^{\alpha_.} \wedge \mathcal{F}_t^{x_0}$ so our assumption implies that $\mathcal{F}_t^{W_.} = \mathcal{F}_t^{\alpha_.}$, for all positive t, after stopping $W_.$ at the explosion time of $\alpha_.$. From this, and equation (6.9), we see that if we set $\bar{M}_t^{df,\alpha} = \mathbf{E}\{M_t^{df,\alpha}|\mathcal{F}^{\alpha_t}\}$ we obtain a martingale with respect to $\mathcal{F}_*^\alpha$ and

$$f(\alpha_t) = \bar{M}_t^{df,\alpha} + \int_0^t \overline{/\!/_s^* \mathcal{B}^V}(f)(\alpha_s)ds \quad (6.13)$$

where $\overline{/\!/_s^* \mathcal{B}^V} = \mathbf{E}\{/\!/_s^* \mathcal{B}^V\}$. Thus by the usual martingale characterisation of Markov processes we see that $\alpha_.$ is Markov with (possibly time dependent) generator $\overline{/\!/_s^* \mathcal{B}^V}$ at time s. However equation (6.12) then implies, for example by [91] Proposition(2.2), Chapter VII, that the generator is given by $(/\!/_s^\sigma)^*(\mathcal{B}^V)$ for arbitrary σ in a set of full measure in $C_{x_0}M$. Thus (ix_0) implies the stochastic holonomy invariance (iix_0).

Conversely if (iix_0) holds, equation (6.10) gives

$$f(\alpha_t) = M_t^{df,\alpha} + \int_0^t \mathcal{B}^V(f)(\alpha_s)ds.$$

Then $M^{df,\alpha}$ is an $\mathcal{F}_*^\alpha$-martingale and again we see that $\alpha_.$ is Markov, with generator $\mathcal{B}^V$. It is therefore independent of $x_.$ giving (ix_0). Moreover, in an obvious notation, if $0 \leqslant s \leqslant t$, by the flow property of parallel translations, on N_{x_0},

$$\mathcal{B}^V = /\!/_t^*(\mathcal{B}^V) = /\!/_s^*(/\!/_t^s)^*(\mathcal{B}^V),$$

and so, almost surely, at all points of N_{x_s} we have

$$(/\!/_t^s)^*(\mathcal{B}^V) = (/\!/_s^*)^{-1}\mathcal{B}^V = \mathcal{B}^V.$$

Since $(/\!/_t^s)^*(\mathcal{B}^V)$ has the same law as $/\!/_{t-s}^*(\mathcal{B}^V)$ and is independent of $\mathcal{F}_s^{x_0}$ this shows that (ii$y$) holds for $p_s^A(x_0, -)$-almost all $y \in M$ for all $s > 0$.

On the other hand (ii$y$) implies that $\mathcal{B}^V$ and $\mathcal{A}^H$ commute on N_y by Lemma 6.2.5 and Theorem 6.0.2. Thus by continuity of $[\mathcal{B}^V, \mathcal{A}^H]$, and the support theorem we see that (iix_0) implies (ivx_0).

Furthermore as in Theorem 6.0.2 we see that (ivx_0) implies that $\mathcal{B}^V$ commutes with basic vector fields at all points over $\overline{\mathcal{D}^0(x_0)}$. From this the holonomy invariance (iii$y$) holds for all $y \in \mathcal{D}^0(x_0)$.

Now assume (iiix_0) and so by Remark 6.2.8(2.) we have (iii$y$) for all $y \in \mathcal{D}^0(x_0)$. Since $/\!/^\sigma_t(u_0)$ stays above $\mathcal{D}^0(x_0)$ for any suitable piecewise smooth σ we find the solution to the martingale problem of $\mathcal{B}^V$ for any point u_0 of N_{x_0} is holonomy invariant at u_0, i.e. along piecewise smooth curves σ in M starting at x_0,

$$P^V_t(f \circ /\!/^\sigma_s -)(u_0) = P^V_t(f)(/\!/^\sigma_s u_0).$$

By Wong-Zakai approximations we see that stochastic holonomy invariance of $P^{\mathcal{B}^V}$ holds over x_0 and hence on taking expectations we get (vx_0). As observed we also get (vy) for all $y \in \mathcal{D}^0(x_0)$ and hence by continuity for all $y \in \overline{\mathcal{D}^0(x_0)}$. Thus (iii$x_0$) implies (v$x_0$).

Finally assuming (vx_0) we can apply Proposition 6.2.2, observing that the proof still holds since it only involves points in $\overline{\mathcal{D}^0(x_0)}$. Differentiating equation (6.5) in s at $s = 0$ gives the stochastic holonomy invariance (ii$y$) for all $y \in \mathcal{D}^0(x_0)$. $\qquad\square$

Remark 6.2.10. From the proof and Theorem 6.0.2 we see that the stochastic completeness of the connection is not needed to ensure that (ivx_0) and (iiix_0) are equivalent.

We can now go further than our Theorem 2.6.1 in extending Hermann's result, Theorem 6.0.1. For this we will need some extra hypoellipticity conditions to deal with the case of non-compact fibres. Take a Hörmander form $\mathcal{A}$ corresponding to a smooth factorisation

$$\sigma^{\mathcal{A}}_x = X(x)X(x)^*$$

with $X(x) \in \mathbb{L}(\mathbf{R}^m : T_x M)$ for $x \in M$ as usual. Let $\mathbf{H}$ denote the usual Cameron-Martin space of finite energy paths $\mathbf{H} = L^{2,1}_0([0,1]; \mathbf{R}^m)$. For $h \in \mathbf{H}$ and $x \in M$ let $\phi^h_t(x), 0 \leqslant t \leqslant 1$ be the solution at time $t \in [0,1]$ to the ordinary differential equation

$$\dot{z}(t) = X(z(t))(\dot{h}) \tag{6.14}$$

with $\phi^h_0(x) = x$. In particular we assume such a solution exists up to time $t = 1$. For each $x \in M$ this gives a smooth mapping $\phi^-_1(x) : \mathbf{H} \to M$, namely $h \mapsto \phi^h_1(x)$. Let $C^{h,x} : E_x \to E_x$ be the **deterministic Malliavin covariance operator**, see [11], given by

$$C^{h,x} = T_h \phi^-_1(x)(T_h \phi^-_1(x))^*.$$

Then $\phi^-_1(x)$ is a submersion in a neighbourhood of h if and only if $C^{h,x}$ is non-degenerate. It is shown in [11] that this condition is independent of the choice of Hörmander form for $\mathcal{A}$, and follows from the standard Hörmander condition that $X^1, \dots, X^m$ and their iterated Lie brackets span $T_x M$ when evaluated at the point x. A more intrinsic formulation of it can be made in terms of the manifold of E-horizontal paths of finite energy, as described in [79].

Theorem 6.2.11. *Consider a smooth map $p : N \to M$ with diffusion operator $\mathcal{B}$ on N over a cohesive diffusion operator $\mathcal{A}$. Suppose that the connection induced by $\mathcal{B}$ is complete. Also assume that $\mathcal{D}^0(x)$ is dense in M for all $x \in M$ and that either the fibres of p are compact or that the solutions to equation (6.14) exist up to time 1 and there exists $h_0 \in \mathbf{H}$ and $x_0 \in M$ such that C^{h_0, x_0} is non-degenerate. Then $p : N \to M$ is a locally trivial bundle.*

If also $\mathcal{B}$ and $\mathcal{A}^H$ commute we can take N_{x_0}, the fibre over x_0, to be the model fibre and choose the local trivialisations

$$\tau : U \times N_{x_0} \to p^{-1}(U)$$

to satisfy

$$\tau(x, -)^* (\mathcal{B}^V | N_x) = \mathcal{B}^V | N_{x_0}.$$

Proof. The local triviality given compactness of the fibres is a special case of Corollary 2.6.3 so we will only consider the other case.

For this set $y = \phi_1^{h_0}(x_0)$. Our assumption on the covariance operator together with the smoothness of $h \mapsto \phi_1^h(x_0)$ implies by the inverse function theorem that there is a neighbourhood U_y of y in M and a smooth immersion $s : U_y \to \mathbf{H}$ with $s(y) = h_0$ and $\phi_1^{s(x)}(x_0) = x$ for $x \in U_y$.

We know from Theorem 2.6.1 that p is a submersion so all its fibres are submanifolds of N. Define $\tau_{U_y} : U_y \times N_{x_0} \to p^{-1}(U_y)$ by using the parallel translation along the curves $\phi_t^{s(x)} : 0 \leqslant t \leqslant 1$ that is:

$$\tau_{U_y}(x, v) = /\!/_1^{\phi_\cdot^{s(x)}}(v), \qquad (x, v) \in (U_y \times N_{x_0}). \tag{6.15}$$

For a general point x of M we can find an $x' \in U_y \cap \mathcal{D}^0(x)$ and argue as in the proof of Theorem 2.6.1 to obtain open neighbourhoods U_x of x in M and $U'_{x'}$ of x' in U_{x_0} and a fibrewise diffeomorphism of $p^{-1}(U'_{x'})$ with $p^{-1}(U_x)$ obtained from parallel translations. This can be composed with a restriction of $\tau_{U_{x_0}}$ to give a trivialisation near x. This proves local triviality. The rest follows directly from Remark 6.2.10 since our trivialisations came from parallel translations. $\square$

Remark 6.2.12. Set

$$\mathbb{G}(\mathcal{B}_{x_0}^V) = \{\alpha \in \text{Diff}(N_{x_0}) : \alpha^*(\mathcal{B}^V | N_{x_0}) = \mathcal{B}^V | N_{x_0}\}. \tag{6.16}$$

Then assuming the commutativity in the theorem we can consider $\mathbb{G}(\mathcal{B}_{x_0}^V)$ as a structure group for our bundle, though unless the fibres of p are compact it is not clear if we have a smooth fibre bundle with this as group in the usual sense, since this requires smoothness into $\mathbb{G}(\mathcal{B}_{x_0}^V)$ of the transition maps between overlapping trivialisations. See the next section and Michor [76] section 13.

Note that elements of $\mathbb{G}(B_{x_0}^V)$ preserve the symbol of $\mathcal{B}^V$ and so if that symbol has constant rank they preserve the inner product induced on the image of $\sigma^{\mathcal{B}^V}$. In particular if $\mathcal{B}^V$ is elliptic they are isometries of the Riemannian structure induced on the fibre N_{x_0}. This is the situation arising from Riemannian submersions as in Hermann's Theorem 6.0.4 and described in detail in Chapter 7 below. The space of isometries of a Riemannian manifold with compact- open topology is well known to form a Lie group, for example see [55]. However there appears to be no detailed proof that the same holds in degenerate cases even when the Hörmander condition holds at each point. When Hörmander's condition holds, the Caratheodory metric on the manifold determines the standard manifold topology, e.g. see [79] Theorem 2.3, which is locally compact, and the group of isometries of a connected locally compact metric space is locally compact in the compact-open topology, see [55], Chapter 1, Theorem 4.7. Thus in this case $\mathbb{G}(B_{x_0}^V)$ will be locally compact.

In general, preserving the possibly degenerate Riemannian structure determined by its symbol will not be enough to characterise $\mathbb{G}(B_{x_0}^V)$. Even in the elliptic case there may be a "drift vector" which needs to be preserved as well and this may lead to $\mathbb{G}(B_{x_0}^V)$ being very small. For example if N_{x_0} is $\mathbf{R}^2$ and $\mathcal{B}^V = \frac{1}{2}\triangle - |x|^2 \frac{\partial}{\partial x^1}$ the group is trivial.

Example 6.2.13. 1. As an example consider the situation described in Section 3.3 of the derivative flow of a stochastic differential equation (3.12) on M acting on the frame bundle GLM to produce a diffusion operator $\mathcal{B}$ on GLM. Assume that M is Riemannian and complete, and that the one-point motions are Brownian motions, so that $\mathcal{A} = \frac{1}{2}\triangle$. Assume also that the connection induced is the Levi-Civita connection. Then if $\mathcal{B}$ and $\mathcal{A}^H$ commute, by Corollary 6.0.3, we see that the co-efficients α and β of $\mathcal{B}^V$ described in Theorem 3.3.2 must be constant along horizontal curves. However as pointed out in the proof of Corollary, 3.4.9 the restriction of $\alpha(u)$ for $u \in GLM$ to anti-symmetric tensors is essentially (one half of) the curvature operator. It follows that the curvature is parallel, $\nabla \mathcal{R} = 0$. In turn this implies, [55] page 303, that M is a local symmetric space and so, if simply connected, a symmetric space. In Section 7.2 we show how such stochastic differential equations arise on any symmetric space. Also from the example in Section 3.3 we see that the standard gradient SDE for Brownian motion on spheres also give derivative flows with this property.

2. For the apparently weaker property of commutativity for the derivative flow $T\xi_t$ of our SDE (3.12) acting directly on the tangent bundle TM, recall first that if the generator $\mathcal{A}$ is cohesive (and even if it just happens that the symbol of $\mathcal{A}$ has constant rank, see [36]), then for $v_t = T\xi_t(v_0)$ and some $v_0 \in T_{x_0}M$ we have the covariant SDE

$$\hat{D}v_t = \breve{\nabla}_{v_t} X dB_t - \frac{1}{2}\mathrm{Ric}^{\#}(v_t)dt + \breve{\nabla}_{v_t}\mathcal{A}dt. \tag{6.17}$$

Here $\breve{\nabla}_{v_t} X$etc. refer to the LW connection of the SDE (3.12) and $\hat{D}$ refers to the adjoint connection, both described in Section 3.3, see also Appendix 9.5.2.

From this we see that if $\mathcal{A}$ is cohesive the process α. defined by $\alpha_t = \hat{/\!/}_t{}^{-1} T\xi_t(v_0)$ satisfies the SDE

$$d\alpha_t = \hat{/\!/}_t{}^{-1}\left(\check{\nabla}_{\hat{/\!/}_t\alpha_t} X dB_t - \frac{1}{2}\check{\mathrm{Ric}}^{\#}(\hat{/\!/}_t\alpha_t)dt + \check{\nabla}_{\hat{/\!/}_t\alpha_t} A dt\right).$$

Suppose also that $A = 0$. We see that α. is independent of $\xi.(x_0)$ if and only if both $\check{\nabla}_- X$ and $\check{\mathrm{Ric}}^{\#}$ are holonomy invariant. If M is Riemannian and the solutions of the SDE are Brownian motions and the induced connection is the Levi-Civita connection we can deduce, as above, using Theorem 6.2.9, that commutativity of the the vertical and horizontal diffusion operators on TM holds only if M is locally symmetric .

Chapter 7

Example: Riemannian Submersions and Symmetric Spaces

7.1 Riemannian Submersions

Recall that when N and M are Riemannian manifolds, a smooth surjection $p : N \to M$ is a *Riemannian submersion* if for each u in N the map $T_u p$ is an orthogonal projection onto $T_{p(u)}M$, i.e. restricted to the orthogonal complement of its kernel, it is an isometry. Note that if $p : N \to M$ is a submersion and M is Riemannian we can choose a Riemannian structure for N which makes p a Riemannian submersion. If a diffusion operator $\mathcal{B}$ on N which has projectible symbol for $p : N \to M$ is also elliptic, its symbol induces Riemannian metrics on N and M for which p becomes a Riemannian submersion. A well-studied situation is when p is a Riemannian submersion and $\mathcal{B}$ is the Laplacian, or $\frac{1}{2}\triangle_N$, on N. The basic geometry of Riemannian submersions was set out by O'Neill in [83]; he ascribes the term 'submersion' to Alfred Gray. In this section we shall mainly be relating the work of Bérard-Bergery & Bourguignon [9], Hermann, [51], Elworthy & Kendall, [33], and Liao, [69], to the discussion above. The book [42] shows the breadth of geometric structures which can be considered in association with Riemannian submersions.

A simple example of a Riemannian submersion is the map $p : \mathbf{R}^n - \{0\} \to \mathbf{R}$ given by $p(x) = |x|$. Then, for $n > 1$, Brownian motion on $\mathbf{R}^n - \{0\}$ is mapped to the Bessel process on $(0, \infty)$ with generator $\mathcal{A} = \frac{1}{2}\frac{d^2}{dx^2} + \frac{n-1}{2x}\frac{d}{dx}$. Thus in this case $\frac{1}{2}\triangle_N$ is projectible but its projection is not $\frac{1}{2}\triangle_M$. The well-known criterion for the latter to hold is that p has *minimal fibres* as we show below. See also [30], and [69].

K.D. Elworthy et al., *The Geometry of Filtering*, Frontiers in Mathematics,
DOI 10.1007/978-3-0346-0176-4_7, © Springer Basel AG 2010

To examine this in more detail we follow Liao [69]. Suppose that p is a Riemannian submersion. The horizontal subbundle on N is just the orthogonal complement of the vertical bundle. Working locally, take an orthonormal family of vector fields $X^1, \ldots, X^n$ in a neighbourhood of a given point x_0 of M. Let $\tilde{X}^1, \ldots, \tilde{X}^n$ be their horizontal lifts to a neighbourhood of some u_0 above x_0, and let $V^1, \ldots, V^p$ be a locally defined orthonormal family of vertical vector fields around u_0. Then near u_0, using the summation convention over $j = 1, \ldots, n$, $\alpha = 1, \ldots, p$, we have

$$\triangle_N = \tilde{X}^j \tilde{X}^j + V^\alpha V^\alpha - \nabla^N_{\tilde{X}^j} \tilde{X}^j - \nabla^N_{V^\alpha} V^\alpha \tag{7.1}$$

while

$$\triangle_M = X^j X^j - \nabla^M_{X^j} X^j. \tag{7.2}$$

Here ∇^M, ∇^N refer to the Levi-Civita connections on M and N, and we are identifying the vector fields with the Lie differentiation in their directions.

Now $\tilde{X}^j \tilde{X}^j$ lies over $X^j X^j$ while $V^\alpha V^\alpha$ is vertical. Also the horizontal component of the sum $\nabla_{V^\alpha} V^\alpha$ at a point $u \in N$ is the trace of the second fundamental form, see the Appendix, Section 9.4, of the fibre $N_{p(u)}$ of p through u, denoted by $T_{V^\alpha} V^\alpha$ in O'Neill's notation, while $\nabla^N_{\tilde{X}^j} \tilde{X}^j$ lies over $\nabla_{X^j} X^j$ by Lemma 1 of [83].

Thus we see that $\frac{1}{2} \triangle_N$ is projectible if and only if the trace of the second fundamental form, trace T, of each fibre $p^{-1}(x)$ is constant along the fibre in the sense of being the horizontal lift of a fixed tangent vector, $2A(x) \in T_x M$. If so, $\frac{1}{2} \triangle_N$ lies over $\frac{1}{2} \triangle_M - A$. In particular $A = 0$, or equivalently p maps Brownian motion to Brownian motion, if and only if p has minimal fibres.

In general to relate to the discussion in Section 2.4, we can set $b^H(u) = -\frac{1}{2}$ trace $T(u)$, with $b(u) = T_u p b^H(u)$ in $T_{p(U)} M$. Let $\triangle^V$ be the vertical operator on N which restricts to the Laplacian on each fibre, and let $\triangle^H$ be the horizontal lift of $\frac{1}{2} \triangle_M$. Our decomposition in Theorem 2.4.6 becomes

$$\frac{1}{2} \triangle_N = \left(\frac{1}{2} \triangle^H - \frac{1}{2} \operatorname{trace} T \right) + \frac{1}{2} \triangle^V \tag{7.3}$$

since the vertical part of $\nabla_{V^\alpha} V^\alpha$ is just $\nabla^V_{V^\alpha} V^\alpha$ where ∇^V refers to the connection on the vertical bundle which restricts to the Levi-Civita of the fibres, and also the vertical part of $\tilde{X}^j \tilde{X}^j$ vanishes because by Lemma 2 of [83] the vertical part of $\tilde{X}^j \tilde{X}^k$ is the vertical part of $\frac{1}{2} [\tilde{X}^j, \tilde{X}^k]$.

7.2   Riemannian Symmetric Spaces

Let K be a Lie group with bi-invariant metric and let M be a Riemannian manifold with a symmetric space structure given by a triple (K, G, σ). This means that there is a smooth left action $K \times M \to M, (k, x) \mapsto L_k(x)$ of K on M by isometries such that, if we fix a point x_0 of M and define $p : K \to M$ by $p(k) = L_k(x_0)$, then p is a Riemannian submersion and a principal bundle with group the subgroup K_{x_0} of K which fixes x_0. Write G for K_{x_0}. Thus M is diffeomorphic to K/G. Moreover if $\mathfrak{g}$ denotes the Lie algebra of G, and $\mathfrak{k}$ that of K, (identified with the tangent spaces at the identity to G and K respectively), there is an orthogonal and ad_G-invariant decomposition

$$\mathfrak{k} = \mathfrak{g} + \mathfrak{m}$$

where $\mathfrak{m}$ is a linear subspace of $T_{\mathrm{id}}K$. Further σ is an involution on K and $\mathfrak{g}$ and $\mathfrak{m}$ are, respectively, the $+1$ and the -1 eigenspaces of the involution on $T_{\mathrm{id}}K$ induced by σ. See Note 7, page 301, of Kobayashi & Nomizu Volume I, [55], for definitions and basic properties, and Volume II, [56], for a detailed treatment.

We shall also let σ denote the involutions induced by σ on $\mathfrak{k}$ and on M, and by differentiation on TM and OM. On M it is an isometry, so it does act on OM. Note that on $T_{x_0}M$ it acts as $v \mapsto -v$.

Since G fixes x_0 the derivative of the left action L_k at x_0 gives a representation of G by isometries of $T_{x_0}M$. The *linear isotropy representation*. We shall assume it to be faithful, i.e. injective. As a consequence the action of K on M is effective, so that K can be considered as a subgroup of the diffeomorphism group of M, and also the action of K on the frame bundle of M is free, i.e the only element of K which fixes a frame is the identity element. See page 187 and the remark on page 198 of [56] for a discussion of this, and how the condition can be avoided. Taking a fixed orthonormal frame $u_0 : \mathbf{R}^n \to T_{x_0}M$, say, at x_0, we can consider G as acting by isometries on $\mathbf{R}^n$ by

$$g \cdot e = u_0^{-1}TL_g u_0(e). \tag{7.4}$$

Let $\rho : G \to O(n)$ denote this representation. We then have the well-known identification of K as a subbundle of the orthonormal frame bundle of M:

Proposition 7.2.1. *Let* $\Phi : K \to OM$ *be defined by* $\Phi(k)(e) = TL_k(u_0e)$ *for* $e \in \mathbf{R}^n$. *Then* Φ *is an injective homomorphism of principle bundles. Moreover* Φ *is equivariant for the actions of* σ *on* K *and* OM.

Proof. To see that Φ is a bundle homomorphism it is only necessary to check that Φ commutes with the actions of G. For this take $e \in \mathbf{R}^n$ and $g \in G$. Then, for $k \in K$,

$$\Phi(k \cdot g)(e) = TL_k TL_g u_0(e)$$
$$= \Phi(k)TL_g u_0(e) = \Phi(k)u_0(g \cdot e)$$

as required. For the equivariance with respect to σ, observe that by definition, $\sigma(L_k x_0) = L_{\sigma(k)} x_0$ so that acting on the frame $\Phi(k)$ we have

$$\sigma(\Phi(k)) = \sigma(TL_k \circ u_0)$$
$$= TL_{\sigma(k)} u_0 = \Phi(\sigma(k)).$$

$\square$

It is easy to see that $p : K \to M$ has totally geodesic fibres. We can therefore take $\mathcal{B} = \frac{1}{2}\triangle^K$ to have $\mathcal{B}$ lying over $\frac{1}{2}\triangle^M$. Moreover in the decomposition of $\mathcal{B}$ the vertical component $\frac{1}{2}\triangle^V$ restricts to one half the Laplacian of G on the fibre $p^{-1}(x_0)$. The induced connection has horizontal subspace $\mathfrak{m}$ at the identity element of K. It is clearly left K-invariant and so $H_k = TL_k[\mathfrak{m}]$ for general $k \in K$. From the equi-variance under the right action of G it is a principle connection: $TR_g[H_k] = H_{kg}$. Since $H_{kg} = TL_k TL_g[\mathfrak{m}] = TR_k TL_k ad_g[\mathfrak{m}]$ this holds because of the ad_G-invariance of $\mathfrak{m}$. This is the *canonical connection*.

The connection on K extends to one on OM as described in Proposition 3.1.3. This is known as the *canonical linear connection*. Since the connection on K is invariant under σ, by the equi-variance of Φ so is the canonical linear connection. As in [56] we have:

Proposition 7.2.2. *The canonical linear connection is the Levi-Civita connection.*

Proof. It is only necessary to check that its torsion T vanishes. By left invariance it is enough to do that at the point x_0. Let $u, v \in T_{x_0}M$. However by invariance under σ we see

$$T(u, v) = \sigma T(\sigma(u), \sigma(v)) = -T(-u, -v) = -T(u, v),$$

as required. □

Let $k_t, t \geqslant 0$ be the canonical Brownian motion on K starting at the identity, id, and let B_t be the Brownian motion on the Euclidean space $\mathfrak{k}$ given by the right flat anti-development:

$$B_t = \int_0^t TR_{k_s}^{-1} d\{k_s\}.$$

Define $\xi_t : M \to M$ by $\xi_t(x) = L_{k_t} x$, for $t \geqslant 0$, $x \in M$.

Proposition 7.2.3. *The diffeomorphism group-valued process $\xi_t, t \geqslant 0$ is the flow of the SDE*

$$dx_t = X(x_t) \circ dB_t$$

where

$$X(x)\alpha = \frac{d}{dt} L_{\exp t\alpha} x|_{t=0}.$$

Proof. Observe that $k.$ satisfies the right invariant SDE,

$$dk_t = TR_{k_t} \circ dB_t,$$

which is p-related to the given SDE on M. □

Remark 7.2.4. The last two propositions relate to the discussion of connections determined by stochastic flows in the next Chapter, and to the discussion about canonical SDE on symmetric spaces in [36]. In [36] it was shown that the connection determined by our SDE is the Levi-Civita connection. In Theorem 8.1.3 below, and in Theorem 3.1 of [34], it is shown that the connection determined by a flow (in this case the canonical linear connection) is the adjoint of that induced by its SDE. This is confirmed in our special case since the adjoint of a Levi-Civita connection is itself.

We can also apply our analysis of the vertical operators and Weitzenböck formulae to our situation, For this it is simplest to assume the symmetric space is *irreducible*. This means that the restricted linear holonomy group of the canonical connection on $p : K \to M$ is irreducible, i.e. for every $g \in G$ there is a null-homotopic loop based at x_0 whose horizontal lift starting at id $\in K$ ends at the point g. The definition in [56] is that $[\mathfrak{m}, \mathfrak{m}]$ acts irreducibly on $\mathfrak{m}$ via the adjoint action, and it is shown there, page 252, that this implies that $\mathfrak{g} = [\mathfrak{m}, \mathfrak{m}]$. As a consequence the linear isotropy representation of G on $T_{x_0}M$ is irreducible, and equivalently so is our representation ρ.

The vertical operators determined by $\mathcal{B}^V$ on the bundles associated to p via our representation ρ and its exterior powers $\wedge^k \rho$ are given in Theorem 3.4.1 by the function $\lambda^{\wedge^k \rho} : K \to \mathcal{L}(\wedge^k \mathbf{R}^n; \wedge^k \mathbf{R}^n)$. By Corollary 3.4.9 and the discussion above they correspond to the Weitzenböck curvatures of the Levi-Civita connection, and so in particular are symmetric. To calculate them using Theorem 3.4.1, first use the fact that $\mathcal{B}^V$ restricts to $\frac{1}{2}\Delta^G$ on $p^{-1}(x_0)$ to represent it as $\frac{1}{2}\sum \mathbf{L}_{A_j^*}\mathbf{L}_{A_j^*}$ for A_j^* as in Section 3.2. The computation in the proof of Corollary 3.4.4 shows that

$$\lambda^{\wedge^k \rho}(u) = -\frac{(n-2)!}{(k-1)!(n-k-1)!}c_{\wedge^k}(u), \tag{7.5}$$

for

$$c_{\wedge^k}(u) = (d\wedge^k)A_l(u) \circ (d\wedge^k)A_l'(u)$$

the Casimir element of our representation $\wedge^k \rho$ of G.

If $\wedge^k \rho$ is irreducible, then $c_{\wedge^k}(u)$ is constant scalar. As remarked in Corollary 3.4.4 this happens when $G = SO(n)$, given our irreducibility hypothesis on the ρ and then it is just $\frac{1}{2}n(n-1)/\frac{n(n-1)\ldots(n-k+1)}{k!}$. Thus for the sphere $S^n(\sqrt{2})$ of radius $\sqrt{2}$, considered as $SO(n+1)/SO(n)$ we have

$$\lambda^{\wedge^k \rho}(u) = -\frac{1}{4}k(n-k). \tag{7.6}$$

7.3 Notes

Intertwining of Laplacians etc. on functions and forms

The result given in Section 7.1, that a Riemannian submersion commutes with the Laplacian if and only if the submersion has minimal fibres, is due to B. Watson, [105]. He also showed that if a map p commutes with the Hodge-Kodaira Laplacian on k-forms, i.e.

$$\Delta p^* \phi = p^* \Delta \phi \qquad\qquad \text{for all } k\text{-forms } \phi$$

for some fixed k with $0 \leqslant k \leqslant \dim M$, then p must be a Riemannian submersion. Goldberg & Ishihara, [48] proved that if this holds for some $k \geqslant 1$ it holds for all k and p is a Riemannian submersion with minimal fibres and integrable horizontal distribution.

Earlier, Watson [106] had proved that a C^2 map $p : N \to M$ between compact oriented Riemannian manifolds commutes with the usual co-differential d^* on k-forms for some $k \geqslant 2$ if and only if p is a totally geodesic Riemannian submersion. Recall that a map is *totally geodesic* if it maps geodesics to geodesics. This corrected earlier work by Lichnerowicz, [71]. If the co-differential is defined to be the trace of the relevant covariant derivative operator, this is a local result so compactness and orientability are not needed.

It was shown by Vilms [104] that a Riemannian submersion is totally geodesic if and only if it has totally geodesic fibres and integrable horizontal distribution. Extending Hermann's result, [51] , Vilms showed that this implies that then, if N is complete so are M and the fibres, and the submersion is a fibre bundle with structure group consisting of isometries of the fibre and an equivariant flat connection. In particular if also M is simply connected the N is just the Riemannian product of M with another Riemannian manifold and p is the projection. It is worth noting the result from the book of Falcitelli, Ianus, and Pastore, [42], that if p is a Riemannian submersion with totally geodesic fibres and N has non-positive sectional curvatures, then so has M and the horizontal distribution is integrable.

For results concerning intertwining of eigenforms of Hodge-Kodaira Laplacians see the book by Gilkey, Leahy, &Park, [46]. For intertwining of certain elliptic pseudo-differential operators, such as powers of the Laplacian, see Furutani [44]. Note that a Riemannian submersion with minimal fibres maps Brownian motion to Brownian motion and so by subordination it will intertwine the powers $(-\Delta)^{\frac{\alpha}{2}}$ for $0 < \alpha \leqslant 2$, [78].

See also Appendix, Section 9.5.2.

Chapter 8

Example: Stochastic Flows

Before analysing stochastic flows by the methods of the previous paragraphs we describe some purely geometric constructions which will enable us to identify the semi-connections which arise in that analysis.

8.1 Semi-connections on the Bundle of Diffeomorphisms

Assume that M is compact. For $r \in \{1, 2, \dots\}$ and $s > r + \dim M/2$, let $\mathcal{D}^s = \mathcal{D}^s(M)$ be the space of diffeomorphisms of M of Sobolev class H^s. See, for example, Ebin-Marsden [28] and Elworthy [30] for the detailed structure of this space. Elements of $\mathcal{D}^s$ are then C^r diffeomorphisms. The space is a topological group under composition, and has a natural Hilbert manifold structure for which the tangent space $T_\theta \mathcal{D}^s$ at $\theta \in \mathcal{D}^s$ can be identified with the space of H^s maps $v : M \to TM$ with $v(x) \in T_{\theta(x)}M$, all $x \in M$. In particular $T_{id}\mathcal{D}^s$ can be identified with the space $H^s\Gamma(TM)$ of H^s vector fields on M. For each $h \in \mathcal{D}^s$ the right translation

$$R_h : \mathcal{D}^s \quad \to \quad \mathcal{D}^s$$
$$R_h(f) \quad = \quad f \circ h$$

is C^∞. However the joint map

$$\mathcal{D}^{s+r} \times \mathcal{D}^s \to \mathcal{D}^s \tag{8.1}$$

is C^r rather than C^∞ for each r in $\{0, 1, 2, \dots\}$.

For $x_0 \in M$ fixed, define $\pi : \mathcal{D}^s \to M$ by

$$\pi(\theta) = \theta(x_0). \tag{8.2}$$

The fibre $\pi^{-1}(y)$ at $y \in M$ is given by: $\{\theta \in \mathcal{D}^s : \theta(x_0) = y\}$. Set $\mathcal{D}^s_{x_0} := \pi^{-1}(x_0)$. Then the elements of $\mathcal{D}^s_{x_0}$ act on the right as C^∞ diffeomorphisms of $\mathcal{D}^s$. We can consider this as giving a principal bundle structure to $\pi : \mathcal{D}^s \to M$ with group $\mathcal{D}^s_{x_0}$, although there is the lack of regularity noted in equation (8.1).

A smooth semi-connection on $\pi : \mathcal{D}^s \to M$ over a subbundle E of TM consists of a family of linear horizontal lift maps $h_\theta : E_{\pi(\theta)} \to T_\theta \mathcal{D}^s$, $\theta \in \mathcal{D}^s$, which is smooth in the sense that it determines a C^∞ section of $\mathcal{L}(\pi^*E; T\mathcal{D}^s) \to \mathcal{D}^s$. In particular we have

$$h_\theta(u) : M \to TM$$

with

$$h_\theta(u)(y) \in T_{\theta(y)}M,$$

$u \in E_{\theta(x_0)}$, $\theta \in \mathcal{D}^s$, $y \in M$.

It is a principal semi-connection if it also has the equivariance property:

$$h_{\theta \circ k}(u)(y) = h_\theta(u)(k(y)) \quad \text{for} \quad k \in \mathcal{D}^s_{x_0}.$$

In this chapter we will only consider principal semi-connections and semi-connections induced by them on associated bundles.

We shall relate principal semi-connections on $\mathcal{D}^s \to M$ to certain reproducing kernel Hilbert spaces. For this let E be a smooth subbundle of TM and $\mathcal{H}$ a Hilbert space which consists of smooth sections of E such that the inclusion $\mathcal{H} \to C^0 \Gamma E$ is continuous (from which comes the continuity into $\mathcal{H}^s \Gamma E$ for all $s > 0$). Such a Hilbert space determines and is determined by its reproducing kernel k, a C^∞ section of the bundle $\mathcal{L}(E^*; E) \to M \times M$ with fibre $\mathcal{L}(E^*_x; E_y)$ at (x, y), see [6]. By definition,

$$k(x, -) = \rho_x^* : E_x^* \to \mathcal{H}$$

where $\rho_x : \mathcal{H} \to E_x$ is the evaluation map at x, and so

$$k(x, y) = \rho_y \rho_x^* : E_x^* \to E_y.$$

Assume $\mathcal{H}$ *spans* E in the sense that for each x in M, $\rho_x : \mathcal{H} \to E_x$ is surjective. It then induces an inner product $\langle , \rangle_x^\mathcal{H}$ on E_x for each x via the isomorphism $\rho_x \rho_x^* : E_x^* \to E_x$.

Using the metric on E the reproducing kernel k induces linear maps

$$k^\#(x, y) : E_x \to E_y, \qquad x, y \in M,$$

with $k^\#(x, x) = \mathrm{id}$.

Proposition 8.1.1. *A Hilbert space $\mathcal{H}$ of smooth sections of a subbundle E of TM which spans E determines a smooth principal semi-connection $h^\mathcal{H}$ on $\pi : \mathcal{D}^s \to M$ over E by*

$$h_\theta^\mathcal{H}(u)(y) = k^\# \Big(\theta(x_0), \theta(y) \Big)(u), \qquad \theta \in \mathcal{D}^s, u \in E_{\theta(x_0)}, y \in M, \qquad (8.3)$$

for $k^\#$ derived from the reproducing kernel of $\mathcal{H}$ as above. In particular the horizontal lift $\tilde{\alpha}$ starting from $\tilde{\alpha}(0) = \mathrm{id}$, of a curve $\alpha : [0, T] \to M$, $\alpha(0) = x_0$ with $\dot{\alpha}(t) \in E_{\alpha(t)}$ for all t, is the flow of the non-autonomous ODE on M,

$$\dot{z}_t = k^\#\left(\alpha(t), z_t\right)\dot{\alpha}(t). \tag{8.4}$$

The mapping $\mathcal{H} \mapsto (h^{\mathcal{H}}, \langle, \rangle^{\mathcal{H}})$ from such Hilbert spaces to principal semi-connections over E and Riemannian metrics on E is injective.

Proof. From the definition of $k^\#$ we see $h_\theta^{\mathcal{H}}(u)(y)$, as given by (8.3), takes values in $T_{\theta(y)}M$, is linear in $u \in E_{\theta(x_0)}$ into $T_\theta \mathcal{D}^s$, and is $\mathcal{D}_{x_0}^s$-invariant. Moreover,

$$T_\theta \pi \circ h_\theta^{\mathcal{H}}(u) = h_\theta^{\mathcal{H}}(u)(x_0) = k^\#\left(\theta(x_0), \theta(x_0)\right)(u) = u$$

for $u \in E_{\theta(x_0)}$ and so $h_\theta^{\mathcal{H}}$ is a 'lift'.

That h is C^∞ as a section of $\mathcal{L}(\pi^* E; T\mathcal{D}^s) \to \mathcal{D}^s$ essentially comes from the smoothness of the map $x \to \rho_x$. More precisely note that for each $r \in \{0, 1, 2, \dots\}$ the composition map

$$T_{id}\mathcal{D}^{r+s} \times \mathcal{D}^s \quad \to \quad T\mathcal{D}^s$$
$$(V, \theta) \quad \mapsto \quad TR_\theta(V)$$

is a C^{r-1} vector bundle map over $\mathcal{D}^s$, being a partial derivative of the composition $\mathcal{D}^{r+s} \times \mathcal{D}^s \to \mathcal{D}^s$. Therefore it induces a C^{r-1} vector bundle map $Z \mapsto TR_\theta \circ Z$, for $Z : E_{\theta(x_0)} \to \mathcal{H}$ and for $\underline{\mathcal{H}}$ the trivial $\mathcal{H}$-bundle over $\mathcal{D}^s$, by composition

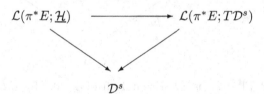

On the other hand $y \mapsto k(y, -)$ can be considered as a C^∞ section of $\mathcal{L}(E; \mathcal{H}) \to M$ and so $\theta \mapsto k(\theta(x_0), -)$ as a C^∞ section of $\mathcal{L}(\pi^* E; \underline{\mathcal{H}})$. This proves the regularity of h.

That the horizontal lift $\tilde{\alpha}$ is the flow of (8.4) is immediate. To see that the claimed injectivity holds, given $h_\theta^{\mathcal{H}}$ observe that (8.3) determines $k^\#$: this is because given any x in M there exists a C^∞ diffeomorphism θ such that $\theta(x_0) = x$ and for such θ,

$$k^\#(x, z)(u) = h_\theta^{\mathcal{H}}(u)(\theta^{-1}z). \tag{8.5}$$

$\square$

Remark 8.1.2. We cannot expect surjectivity of the map $\mathcal{H} \to h^{\mathcal{H}}$ into the space of principal semi-connections on $\pi : \mathcal{D}^s \to M$. Indeed for $k^{\#}$ given by (8.5) to correspond to the reproducing kernel for some Hilbert space of sections of E we need some specific conditions.

1) $h_{\theta}^{\mathcal{H}}(u)(y) \in E_{\theta(y)}$ for $u \in E_{\theta(x_0)}, y \in M$, and a metric $\langle , \rangle$ on E with respect to which the following holds:

2) for $x, y \in M$,

$$k^{\#}(x, y) = \left(k^{\#}(y, x) \right)^{*},$$

3) for any finite set S of points of M and $\{\xi_a\} \in E_a, a \in S,$

$$\sum \left\langle k^{\#}(a, b)\xi_a, \xi_b \right\rangle \geqslant 0.$$

For each frame $u_0 : \mathbf{R}^n \to T_{x_0}M$ there is a homomorphism of principal bundles

$$\begin{aligned} \Psi^{u_0} : \quad \mathcal{D}^s \quad &\to \quad GLM \\ \theta \quad &\mapsto \quad T_{x_0}\theta \circ u_0. \end{aligned} \qquad (8.6)$$

As with connections such a homeomorphism maps a principal semi-connection on $\mathcal{D}^s$ over E to one on GLM. The horizontal lift maps are related by

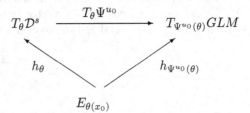

and if $\tilde{\alpha} : [0, T] \to \mathcal{D}^s$ is a horizontal lift of $\alpha : [0, T] \to M$, then

$$\Psi^{u_0}(\tilde{\alpha}(t)) = T_{x_0}\tilde{\alpha}(t) \circ u_0, \qquad 0 \leqslant t \leqslant T$$

is a horizontal lift of α to GLM.

Theorem 8.1.3. *Let $h^{\mathcal{H}}$ be the semi-connection on $\pi : \mathcal{D}^s \to M$ over E determined by some $\mathcal{H}$ as in Proposition 8.1.1. Then the semi-connection induced on GLM, and so on TM, by the homeomorphism Ψ^{u_0} is the adjoint $\hat{\nabla}$ of the metric connection which is projected on $(E, \langle , \rangle^{\mathcal{H}})$ by the evaluation map $(x, e) \mapsto \rho_x(e)$ from $M \times \mathcal{H} \to E$, c.f. (1.1.10) in [36]. In particular every semi-connection on TM with metric adjoint connection arises this way from some, even finite dimensional, choice of $\mathcal{H}$.*

Proof. Let $\alpha : [0, T] \to M$ be a C^1 curve with $\dot{\alpha}(t) \in E_{\alpha(t)}$ for each t. By Proposition 8.1.1 its horizontal lift $\tilde{\alpha}$ to $\mathcal{D}^s$ starting from $\theta \in \pi^{-1}(\alpha(0))$ is the solution to

$$\frac{d\tilde{\alpha}}{dt} = k^{\#}\left(\tilde{\alpha}(t)(x_0), \tilde{\alpha}(t) - \right)\dot{\alpha}(t), \tag{8.7}$$

$$\tilde{\alpha}(0) = \theta. \tag{8.8}$$

The horizontal lift to GLM is $t \mapsto T_{x_0}\tilde{\alpha}(t) \circ u_0$ and to TM through $v_0 \in T_{\theta(x_0)}M$, *i.e.* the parallel translation $\{/\!/_t(v_0) : 0 \leqslant t \leqslant T\}$ of v_0 along α, is given by

$$/\!/_t(v_0) = T_{x_0}\tilde{\alpha}(t) \circ (T_{x_0}\theta)^{-1}(v_0) = T_{\alpha(0)}\left(\tilde{\alpha}(t) \circ \theta^{-1}\right)(v_0).$$

However this is $T_{\alpha(0)}\pi_t(v_0)$ for $\{\pi_t : 0 \leqslant t \leqslant T\}$ the solution flow of

$$\frac{dz_t}{dt} = k^{\#}\left(\alpha(t), z(t)\right)\dot{\alpha}(t)$$

which by Lemma 1.3.4 of [36] is the parallel translation of the adjoint of the associated connection (in [36] $k^{\#}$ is denoted by k).

The fact that all such semi-connections on TM arise from some finite dimensional $\mathcal{H}$ comes from Narasimhan-Ramanan [80] as described in [36], or more directly from Quillen [89] □

8.2 Semi-connections Induced by Stochastic Flows

From Baxendale [7] we know that a C^{∞} stochastic flow $\{\xi_t : t \geqslant 0\}$ on M, i.e. a Wiener process on $\mathcal{D}^{\infty} := \cap_s \mathcal{D}^s$, can be considered as the solution flow of a stochastic differential equation on M driven by a possibly infinite dimensional noise. Its one-point motions form a diffusion process on M with generator $\mathcal{A}$, say. The noise comes from the Brownian motion $\{W_t : t \geqslant 0\}$ on $\mathcal{H}^s\Gamma(TM)$ determined by a Gaussian measure γ on $\mathcal{H}^s\Gamma(TM)$. (In our C^{∞} case they lie on $\mathcal{H}^{\infty}(TM) := \cap_s \mathcal{H}^s\Gamma(TM)$.) We will take γ to be mean zero and so we may have a drift A in $\mathcal{H}^{\infty}(TM)$. The stochastic flow $\{\xi_t : t \geqslant 0\}$ can then be taken to be the solution of the right invariant stochastic differential equation on $\mathcal{D}^s$,

$$d\theta_t = TR_{\theta_t} \circ dW_t + TR_{\theta_t}(A)dt \tag{8.9}$$

with ξ_0 the identity map id. In particular it determines a right invariant generator $\mathcal{B}$ on $\mathcal{D}^s$.

For fixed x_0 in M the one-point motion $x_t := \xi_t(x_0)$ solves

$$dx_t = \circ dW_t(x_t) + A(x_t)dt. \tag{8.10}$$

We can write (8.10) as

$$dx_t = \rho_{x_t} \circ dW_t + A(x_t)dt. \tag{8.11}$$

Thus $\pi(\xi_t) = \xi_t(x_0) = x_t$. For a map θ in $\mathcal{D}^s$, the solution $\xi_t \circ \theta$ to (8.9) starting at θ has $\pi(\xi_t \circ \theta) = \xi_t(\pi(\theta))$, the solution to (8.11) starting from $\pi(\theta)$, and we see that the diffusions are π-related (c.f. [30]), and $\mathcal{A}$ and $\mathcal{B}$ are intertwined by π.

The measure γ corresponds to a reproducing kernel Hilbert space, H_γ say, or equivalently to an abstract Wiener space structure $i : H_\gamma \to \mathcal{H}^s\Gamma(TM)$ with i the inclusion (although i may not have a dense image). Then

$$\sigma_\theta^\mathcal{B} : (T_\theta \mathcal{D}^s)^* \to T_\theta \mathcal{D}^s$$

is right invariant and determined at $\theta = id$ by the canonical isomorphism $H_\gamma^* \simeq H_\gamma$ through the usual map $j = i^*$,

$$(\mathcal{H}^s\Gamma(TM))^* \overset{j}{\hookrightarrow} H_\gamma^* \simeq H_\gamma \overset{i}{\hookrightarrow} \mathcal{H}^s\Gamma(TM),$$

i.e.

$$\sigma_{id}^\mathcal{B} = i \circ j.$$

This shows that H_γ is the image of $\sigma_{id}^\mathcal{B}$ with induced metric. In this situation our cohesiveness condition on $\mathcal{A}$ becomes the assumption that there is a C^∞ subbundle E of TM such that H_γ consists of sections of E and spans E, and A is a section of E. Let $\langle , \rangle_y$ be the inner product on E_y induced by H_γ.

The reproducing kernel k of H_γ is the covariance of γ and:

$$k^\#(x,y)v = \int_{U \in \mathcal{H}^s\Gamma(E)} \Big\langle U(x), v \Big\rangle_x U(y)\, d\gamma(U), \qquad v \in E_x;\ x,y \in M.$$

Analogously to Lemma 2.3.1 we have the commutative diagram

$$
\begin{array}{ccccc}
(T_\theta\mathcal{D}^s)^* & \xrightarrow{\;\; j \circ (T\mathcal{R}_\theta)^* \;\;} & H_\gamma & \xrightarrow{\;\; T\mathcal{R}_\theta \circ i \;\;} & T_\theta\mathcal{D}^s \\[2mm]
\big\uparrow{\scriptstyle (T_\theta\pi)^*} & & \big\uparrow{\scriptstyle \ell_\theta} & & \big\downarrow{\scriptstyle T_\theta\pi = \rho_{x_0}} \\[2mm]
T^*_{\theta(x_s)}M \to E^*_{\theta(x_0)} & \xrightarrow[\;\; k(\theta(x_0),-) \;\;]{\;\; \rho_{\theta(x_0)} \;\;} & H_\gamma & \xrightarrow{\;\; \rho_{\theta(x_0)} \;\;} & E_{\theta(x_0)} \hookrightarrow T_{x_0}M
\end{array}
$$

with ℓ_θ uniquely determined under the extra condition

$$\ker \ell_\theta = \ker \rho_{\theta(x_0)}.$$

Writing $K : M \to \mathcal{L}(H_\gamma; H_\gamma)$ for the map giving the projection $K(x)$ of H_γ onto $\ker \rho_x$ for each x in M, and letting $K^\perp(x)$ be the projection onto $[\ker \rho_x]^\perp$, we have

$$\ell_\theta = K^\perp\big(\theta(x_0)\big),$$

(agreeing with the note following Lemma 2.3.1), and so

$$\ell_\theta(U) = k^\# \Big(\theta(x_0), - \Big) U(\theta(x_0)), \qquad U \in \mathcal{H}_\gamma.$$

Note that the formula

$$K^\perp(y)(U) = k^\#(y, -)U(y)$$

for U in $\mathcal{H}_\gamma$ determines an extension $K^\perp(y) : \Gamma E \to \mathcal{H}_\gamma$. We then define $K(y)U = U - K^\perp(y)U$. Note that $\rho_y(K(y)U) = 0$ for all U in ΓE.

The horizontal lift map determined by $\mathcal{B}$ as in Proposition 2.1.2 is therefore given by

$$h_\theta : E_{\theta(x_0)} \to T\mathcal{R}_\theta(\mathcal{H}_\gamma) \qquad \subset \quad T_\theta \mathcal{D}^s$$
$$h_\theta(u) = T\mathcal{R}_\theta \, \ell_\theta \Big[k^\#(\theta(x_0), -)u \Big], \tag{8.12}$$

for $\theta \in \mathcal{D}^s$. Consequently

$$h_\theta(u)(y) = k^\# \Big(\theta(x_0), \theta(y) \Big)(u). \tag{8.13}$$

Comparing this with formula (8.3) we have

Proposition 8.2.1. *The semi-connection h determined on $\pi : \mathcal{D}^s \to M$ by the equivariant diffusion operator $\mathcal{B}$ is just that given by the reproducing kernel Hilbert space $\mathcal{H}_\gamma$ of the stochastic flow which determines $\mathcal{B}$, i.e.*

$$h = h^{\mathcal{H}_\gamma}.$$

The horizontal lift $\{\tilde{x}_t : t \geq 0\}$ of the one-point motion $\{x_t : t \geq 0\}$ with $\tilde{x}_0 = \mathrm{id}$ is the solution to

$$d\tilde{x}_t = k^\# \Big(\tilde{x}_t(x_0), \tilde{x}_t - \Big) \circ dx_t; \tag{8.14}$$

which in a more revealing notation is:

$$d\tilde{x}_t = T\mathcal{R}_{\tilde{x}_t} \Big(K^\perp(\tilde{x}_t(x_0)) \circ dW_t \Big) + T\mathcal{R}_{\tilde{x}_t} \Big(K^\perp(\tilde{x}_t(x_0))A \Big) \, dt. \tag{8.15}$$

Equivalently $\{\tilde{x}_t : t \geq 0\}$ can be considered as the solution flow of the non-autonomous stochastic differential equation on M,

$$dy_t \quad = \quad k^\# \big(x_t, y_t \big) \circ dx_t,$$

i.e.

$$dy_t = \Big(K^\perp(x_t) \circ dW_t \Big)(y_t) + K^\perp(x_t)(A)(y_t) \, dt. \tag{8.16}$$

The standard fact that the solution to such an equation as (8.16) starting at x_0 is just $\{x_t : t \geq 0\}$, i.e. that $\tilde{x}_t(x_0) = x_t$ reflects the fact that $\tilde{x}$. is a lift of x.. The lift through $\phi \in \mathcal{D}^s_{x_0}$ is just $\{\tilde{x}_t \circ \phi : t \geq 0\}$.

Remark 8.2.2. If our solution flow is that of an SDE,

$$dx_t = X(x_t) \circ dB_t + A(x_t)dt$$

for $X(x) : \mathbf{R}^m \to TM$ arising, for example, from a Hörmander form representation of $\mathcal{A}$ as in §4.7 above, the relationships with the notation in this section is as follows: $H_\gamma = \{X(\cdot)e : e \in \mathbf{R}^m\}$ with inner product induced by the surjection $\mathbf{R}^m \to H_\gamma$. If $Y_x = [X(x)|_{\ker X(x)^\perp}]^{-1}$, then $k^\#(y, -) : E_y \to H_\gamma$ is

$$k^\#(y, -)u = X(-)Y_y(u), \qquad u \in E_y.$$

Also $K^\perp(y) : \Gamma E \to H_\gamma$ is $K^\perp(y)U = X(-)Y_y(U(y))$.

Remark 8.2.3. The reproducing kernel Hilbert space H_γ determines the stochastic flow and so by the injectivity part of Proposition 8.1.1 the semi-connection together with the generator $\mathcal{A}$ of the one-point motion determines the flow, or equivalently the operator $\mathcal{B}$. This is because the symbol of $\mathcal{A}$ again gives the metric on E which together with the semi-connection determines H_γ by Proposition 8.1.1. The generator $\mathcal{A}$ then determines the drift A. A consequence is that the horizontal lift $\mathcal{A}^H$ of $\mathcal{A}$ to $\mathcal{D}^s$ determines the flow (and hence $\mathcal{B}$, so $\mathcal{B}^V$ really is redundant).

To see this directly note that given any cohesive $\mathcal{A}$ on M and $\mathcal{D}^s_{x_0}$-equivariant $\mathcal{A}^H$ on $\mathcal{D}^s$ over $\mathcal{A}$, with no vertical part, there is at most one vertical $\mathcal{B}^V$ such that $\mathcal{A}^H + \mathcal{B}^V$ is right invariant. This results from the following lemma.

Lemma 8.2.4. *Suppose $\mathcal{B}^1$ is a diffusion operator on $\mathcal{D}^s$ which is vertical and right invariant, then $\mathcal{B}^1 = 0$.*

Proof. By Remark 1.3.4 (i) the image $\mathcal{E}_\theta$, say, of $\sigma^{\mathcal{B}^1}_\theta$ lies in $VT_\theta\mathcal{D}^s$ for $\theta \in \mathcal{D}^s$ and so if $V \in \mathcal{E}_\theta$. On the other hand, by right invariance $\mathcal{E}_\theta = TR_\theta(\mathcal{E}_{\mathrm{id}})$. Therefore if $V \in \mathcal{E}_{\mathrm{id}}$, then $V(\theta(x_0)) = TR_\theta(V)(x_0) = 0$ all $\theta \in \mathcal{D}^s$ and so $V \equiv 0$. Thus $\mathcal{E}_{\mathrm{id}} = \{0\}$ and by right invariance, $\mathcal{B}^1$ must be first order and so given by some vector field Z on $\mathcal{D}^s$. But Z must be vertical and right invariant, so again we see $Z \equiv 0$. $\qquad\qquad\square$

Proposition 3.1.3 applies to the homomorphism $\Psi^{u_0} : \mathcal{D}^s \to GL(M)$ of (8.6). From this and Theorem 8.1.3 we see that the semi-connection ∇ on GLM determined by the generator of the derivative flow in §3.3 is the adjoint $\hat{\nabla}$ of the connection $\breve{\nabla}$, so giving an alternative proof of the first part of Theorem 3.3.2 above. Proposition 3.1.3 also gives a relationship between the curvature and holonomy group of $\hat{\nabla}$ and those of the connection induced by the flow on $\mathcal{D}^s \overset{\rho_{x_0}}{\to} M$.

We can summarize our decomposition results as applied to these stochastic flows in the following theorem. The skew-product decomposition was already described in [34] for the case of solution flows of SDE of the form (4.20), and in particular with finite dimensional noise: however the difference is essentially that of notation, see Remark 8.2.2 above.

Theorem 8.2.5. *Let* $\{\xi_t : t \geqslant 0\}$ *be a* C^∞ *stochastic flow on a compact manifold* M. *Let* $\mathcal{A}$ *be the generator of the one-point motion on* M *and* $\mathcal{B}$ *the generator of the right invariant diffusion on* $\mathcal{D}^s$ *determined by* $\{\xi_t : t \geqslant 0\}$. *Assume* $\mathcal{A}$ *is strongly cohesive. Then there is a unique decomposition* $\mathcal{B} = A^H + \mathcal{B}^V$ *for* A^H *a diffusion operator which has no vertical part in the sense of Definition 2.2.3 and* $\mathcal{B}^V$ *a diffusion operator which is along the fibres of* ρ_{x_0}, *both invariant under the right action of* $\mathcal{D}^s_{x_0}$. *The diffusion process* $\{\theta_t : t \geqslant 0\}$ *and* $\{\phi_t : t \geqslant 0\}$ *corresponding to* A^H *and* $\mathcal{B}^V$ *respectively can be represented as solutions to*

$$d\theta_t = TR_{\theta_t}\left(K^\perp(\theta_t(x_0)) \circ dW_t\right) + TR_{\theta_t}\left(K^\perp(\theta_t(x_0))A\right) dt \qquad (8.17)$$

and

$$d\phi_t = TR_{\phi_t}\left(K(z_0) \circ dW_t\right) + TR_{\phi_t}\left(K(z_0)A\right) dt \qquad (8.18)$$

for $z_0 = \phi_0(x_0) = \phi_t(x_0)$. *There is the corresponding skew-product decomposition of the given stochastic flow*

$$\xi_t = \tilde{x}_t g_t^{x.}, 0 \leqslant t < \infty$$

where $\{\tilde{x}_t : t \geqslant 0\}$ *is the horizontal lift of the one-point motion* $\{\xi_t(x_0) : t \geqslant 0\}$ *with* $\tilde{x}_0 = \mathrm{id}_M$ *and for* $\mathbf{P}^\mathcal{A}_{x_0}$-*almost all* $\sigma : [0, \infty) \to M$, $\{g_t^\sigma : t \geqslant 0\}$ *is a* $\mathcal{D}^s_{x_0}$-*valued process independent of* $\{\tilde{x}_t : t \geqslant 0\}$ *and satisfying*

$$dg_t^\sigma = T\tilde{\sigma}_t^{-1}\rho(\tilde{\sigma}_t g_t^\sigma -)\left(K(\sigma_t) \circ dW_t\right) + T\tilde{\sigma}_t^{-1}\rho(\tilde{\sigma}_t g_t^\sigma -)\left(K(\sigma_t)A\right) dt,$$

$$\tilde{g}_0^\sigma = \mathrm{id}_M$$

where $\tilde{\sigma}$ *is the horizontal lift of* σ *to* $\mathcal{D}^s$ *with* $\tilde{\sigma}_0 = \mathrm{id}_M$.

Remark 8.2.6. We could rewrite the terms such as $K(\sigma_t) \circ dW_t$ and $K^\perp(\sigma_t) \circ dW_t$ above in tems of Itô differentials. As in [36], see Section 2.3.2, these can be written as

$$K(\sigma_t)dW_t = /\!\!/_t(\sigma.) \, d\beta_t$$

$$K^\perp(\sigma_t)dW_t = /\!\!/_t(\sigma.) \, d\tilde{B}_t$$

where $/\!\!/_t(\sigma.) : H_\gamma \to H_\gamma$, $0 \leqslant t < \infty$, is a family of orthogonal transformations mapping $\ker \rho_{x_0} \to \ker \rho_{\sigma_t}$ defined for $\mathbf{P}^\mathcal{A}_{x_0}$-almost all $\sigma : [0, \infty) \to M$ and $\{\beta_t : t \geqslant 0\}$, $\{\tilde{B}_t : t \geqslant 0\}$ are independent Brownian motions, (β_t could be cylindrical), on $\ker \rho_{x_0}$ and $[\ker \rho_{x_0}]^\perp$ respectively.

Proof. Our general result gives the decomposition $\mathcal{B} = A^H + \mathcal{B}^V$ into horizontal and vertical parts. We have just proved the representation (8.17) for A^H. To show that $\mathcal{B} - A^H$ corresponds to (8.18) take an orthonormal base $\{X^j\}$ for H_γ. Then, on a suitable domain,

$$\mathcal{B} = \frac{1}{2}\sum_j \mathbf{L}_{X^j}\mathbf{L}_{X^j} + L_\mathbb{A}, \qquad (8.19)$$

for $\mathbb{X}^j(\theta) = TR_\theta(X^j)$ and $\mathbb{A} = TR_\theta(A)$, while, by (8.17),

$$A^H = \frac{1}{2}\sum_j \mathbf{L}_{\mathbb{Y}^j}\mathbf{L}_{\mathbb{Y}^j} + \mathbf{L}_{\mathbb{B}} \tag{8.20}$$

for $\mathbb{Y}^j(\theta) = TR_\theta\left(K^\perp(\theta(x_0))X^j\right)$, $\mathbb{B} = TR_\theta(K^\perp(\theta(x_0))A)$.
   Define vector fields $\mathbb{Z}^j$, $\mathcal{C}$ on $\mathcal{D}^s$ by

$$\mathbb{Z}^j(\phi) = TR_\phi\left(K(\phi(x_0))X^j\right), \qquad \text{and}$$
$$\mathcal{C}(\phi) = TR_\phi\left(K(\phi(x_0))A\right), \qquad \text{for } \phi \in \mathcal{D}^s.$$

Then $\mathbb{A} = \mathbb{B} + \mathcal{C}$ and $\mathbb{X}^j = \mathbb{Y}^j + \mathbb{Z}^j$ each j. Moreover

$$\sum_j \mathbf{L}_{\mathbb{Y}^j}\mathbf{L}_{\mathbb{Z}^j} + \sum_j \mathbf{L}_{\mathbb{Z}^j}\mathbf{L}_{\mathbb{Y}^j} = 0$$

by Lemma 8.2.7 below. This shows that

$$\mathcal{B}^V = \frac{1}{2}\sum_j \mathbf{L}_{\mathbb{Z}^j}\mathbf{L}_{\mathbb{Z}^j} + \mathbf{L}_{\mathcal{C}}. \tag{8.21}$$

Thus the diffusion process from ϕ_0 corresponding to $\mathcal{B}^V$ can be represented by the solution to

$$d\phi_t = TR_{\phi_t}\left(K(\phi_t(x_0) \circ dW_t)\right) + TR_{\phi_t}\left(K(\phi_t(x_0)A)\right)dt. \tag{8.22}$$

If we set $z_t = \rho_{x_0}(\phi_t) = \phi_t(x_0)$, we obtain, via Itô's formula,

$$dz_t = \rho_{z_t}\left(K(z_t) \circ dW_t\right) + \rho_{z_t}\left(K(z_t)A\right)\ dt,$$

i.e. $dz_t = 0$. Thus $\phi_t(x_0) = z_0$ and (8.18) holds.
   The skew-product formula is seen to hold by calculating the stochastic differential of $\tilde{x}_t g_t^{\tilde{x}}$ using (8.15) to see that it satisfies the SDE (8.9) for $\{\xi_t : t \geqslant 0\}$. $\square$

Lemma 8.2.7.
$$\sum_j \mathbf{L}_{\mathbb{Y}^j}\mathbf{L}_{\mathbb{Z}^j} + \mathbf{L}_{\mathbb{Z}^j}\mathbf{L}_{\mathbb{Y}^j} = 0.$$

Proof. Since, for fixed θ, we can choose our basis $\{X^j\}$, such that either $\mathbb{Y}^j(\theta) = 0$ or $\mathbb{X}^j(\theta) = 0$, and since for $f : \mathcal{D}^s \to \mathbf{R}$ we can write

$$df\left(\mathbb{Z}^j(\theta)\right) = \left(df \circ TR_\theta\right)\left(K(\theta(x_0))X^j\right)$$

and

$$df\left(\mathbb{Y}^j(\theta)\right) = (df \circ TR_\theta)\left(K^\perp(\theta(x_0))X^j\right), \qquad \theta \in \mathcal{D}^s,$$

it suffices to show that

$$\sum_j \left\{ (dK^\perp)_{\theta(x_0)} \Big(\mathbb{Z}^j(\theta)(x_0) \Big) X^j + (dK)_{\theta(x_0)} \left(\mathbb{Y}^j(\theta)(x_0) \right) X^j \right\} = 0, \qquad (8.23)$$

for all $\theta \in \mathcal{D}^s$.

Now $K^\perp(y)K(y) = 0$ for all $y \in M$. Therefore

$$(dK^\perp)_y(v)K(y) + K^\perp(y)(dK)_y(v) = 0, \qquad \forall v \in T_x M, x \in M.$$

Writing

$$X^j = K\big(\theta(x_0)\big)X^j + K^\perp\big(\theta(x_0)\big)X^j$$

this reduces the right-hand side of (8.23) to

$$\sum_j (dK^\perp)_{\theta(x_0)} \Big(\mathbb{Z}^j(\theta)(x_0) \Big) \Big(K^\perp(\theta(x_0))X^j \Big)$$

$$+ (dK)_{\theta(x_0)} \Big(\mathbb{Y}^j(\theta)(x_0) \Big) \Big(K(\theta(x_0))X^j \Big) = 0;$$

with our choice of basis this clearly vanishes, as required. $\qquad \square$

8.3 Semi-connections on Natural Bundles

Our bundle $\pi : \mathrm{Diff}\, M \to M$ can be considered as a universal natural bundle over M, and a connection on it induces a connection on each natural bundle over M. Natural bundles are discussed in Kolar-Michor-Slovak [57]), they include bundles such as jet bundles as well as the standard tensor bundles. For example let G_n^r be the Lie group of r-jets of diffeomorphisms $\theta : \mathbf{R}^n \to \mathbf{R}^n$ with $\theta(0) = 0$ for a positive integer r. The diffeomorphisms need only be locally defined so θ can be taken to be a diffeomorphism of an open set U_θ of $\mathbf{R}^n$ onto an open subset of R^n, mapping 0 to 0. Its r-jet at 0, denoted by $j_0^r(\theta)$, is the equivalence class of θ where C^∞ maps $\theta^i : U_{\theta^i} \to \mathbf{R}^n$ for $i = 1, 2$ are equivalent if $\theta^1(0) = \theta^2(0)$ and their first r derivatives at 0 are the same. Using local co-ordinates there is a corresponding definition for r-jets of smooth maps $\phi : U_\phi \to M$ where U_ϕ is an open neighbourhood of some given point x_0 of a manifold N. The equivalence class is then denoted by $j_{x_0}^r(\theta)$.

An **r-th order frame** u at a point x of M is the r-jet at 0 of some $\Psi : U \to M$ which maps an open set U of $\mathbf{R}^n$ diffeomorphically onto an open subset of M with $0 \in U$ and $\Psi(0) = x$. Clearly G_n^r acts on the right of such jets, by composition. From this we can define the r-th order frame bundle $G_n^r M$ of M with group G_n^r as the collection of all r-frames at all points of M.

If we fix an r-th order frame u_0 at x_0 we obtain a homomorphism of principal bundles

$$\Psi^{u_0} : \mathcal{D}^s \to G_n^r M$$

$$\theta \mapsto j_{x_0}^r(\theta) \circ u_0,$$

as for GLM (which is the case $r = 1$) with associated group homomorphism $\mathcal{D}_{x_0}^s \to G_n$ given by $\theta \to u_0^{-1} \circ j_{x_0}^r(\theta) \circ u_0$. As for the case $r = 1$ there is a diffusion operator induced by the flow on $G_n^r M$ and we are in the situation of Theorem 8.1.3. The behaviour of the flow induced on $G_n^2 M$ is essentially that of $j_{x_0}^2(\xi_t)$ and so relevant to the effect on the curvature of submanifolds of M as they are moved by the flow, *e.g.* see Cranston-Le Jan [20], Lemaire [67].

Alternatively rather having to choose some u_0 we see that $G_n^r M$ is (weakly) associated to $\pi : \mathcal{D}^s \to M$ by taking the action of $\mathcal{D}_{x_0}^s$ on $(G_x^r M)_{x_0}$ by

$$(\theta, \alpha) \mapsto j_{x_0}^r(\theta) \circ \alpha.$$

As a geometrical conclusion we can observe:

Theorem 8.3.1. *Any classifying bundle homomorphism*

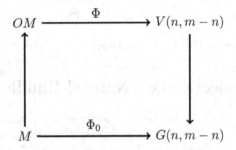

for the tangent bundle to a compact Riemannian manifold M, (where $G(n, m - n)$ is the Grassmannian of n-planes in $\mathbf{R}^m$ and $V(n, m - n)$ the corresponding Stiefel manifold) induces not only a metric connection on TM as the pull-back of Narasimhan and Ramanan's universal connection ϖ_U, but also a connection on $\Pi : \mathcal{D}^s \to M$. The latter induces a connection on each natural bundle over M to form a consistent family; that induced on the tangent bundle is the adjoint of $\Phi^*(\varpi_U)$. The above also holds with smooth stochastic flows replacing classifying bundle homomorphisms, and the resulting map from stochastic flows to connections on $\pi : \mathcal{D}^s \to M$ is injective.

Proof. It is only necessary to observe that Φ determines and is determined by a surjective vector bundle map $X : M \times \mathbf{R}^m \to TM$ (*e.g.* see [36], Appendix 1). This in turn determines a Hilbert space $\mathcal{H}$ of sections of TM as in Remark 8.2.2, so we can apply Proposition 8.1.1 and Theorem 8.1.3. $\square$

Some of the conclusions of Theorem 8.3.1 are explored further in [39].

Remark 8.3.2. This injectivity result in Theorem 8.3.1 implies that all properties of the flow can, at least theoretically, be obtainable from the induced connection on $\mathcal{D}^s$.

Flows on Non-compact Manifolds

In general if M is not compact we will not be able to use the Hilbert manifolds $\mathcal{D}^s$, or other Banach manifolds without growth conditions on the coefficients of our flow. One possibility could be to use the space Diff M of all smooth diffeomorphisms using the Frölicher-Kriegl differential calculus as in Michor [76]. In order to do any stochastic calculus we would have to localize and use Hilbert manifolds (or possibly rough path theory). The geometric structures would nevertheless be on Diff M. This was essentially what was happening in the compact case. However it is useful to include partial flows of stochastic differential equations which are not strongly complete, see Kunita [59] or Elworthy [30]. For the partial solution flow $\{\xi_t : t < \tau\}$ of an SDE as in Remark 8.2.2 we obtain the decomposition in Theorem 8.2.5 but now only for $\xi_t(x)$ defined for $t < \tau(x, -)$. This can be proved from the compact versions by localization as in Carverhill-Elworthy [17] or Elworthy [30].

Chapter 9

Appendices

9.1 Girsanov-Maruyama-Cameron-Martin Theorem

To apply the Girsanov-Maruyama theorem it is often thought necessary to verify some condition such as Novikov's condition to ensure that the exponential (local) martingale arising as Radon-Nikodym derivative is a true martingale. In fact for conservative diffusions this is automatic, and we give a proof of this fact here since it is not widely appreciated. The proof is along the lines of that given for elliptic diffusions in [30] but with the uniqueness of the martingale problem replacing the uniqueness of minimal semi-groups used in [30]. See also [65]. On the way we relate the expectation of the exponential local martingale to the probability of explosion of the trajectories of the associated diffusion process: a special case of this appeared in [75]. Let $\mathcal{B}$ be a conservative diffusion operator on a smooth manifold N. For fixed $T > 0$ and $y_0 \in N$ let $\mathbf{P}_{y_0}^{\mathcal{B}}$ denote the solution to the martingale problem for $\mathcal{B}$ on $C_{y_0}([0,T];N^+)$ and let $\{P_t^{\mathcal{B}}\}_t$ denote the corresponding Markov semi-group acting on bounded measurable functions. Choose an increasing sequence $\{D_n\}_n$ of connected open domains in N with smooth boundary which cover N. Let τ_n denote the first exit time from D_n. For $n = 1,\ldots$ and $y_0 \in D_n$ let $\mathbf{P}_{y_0}^{\mathcal{B},n}$ be the probability measure on $C_{y_0}([0,T];D_n^+)$ giving the solution to the martingale problem for the restriction of $\mathcal{B}$ to D_n, and let $\{P_t^{\mathcal{B},n}\}_t$ be its Markov semi-group. It corresponds to Dirichlet boundary conditions on D_n.

Remark 9.1.1. The measure $\mathbf{P}_{y_0}^{\mathcal{B},n}$ is the push-forward of $\mathbf{P}_{y_0}^{\mathcal{B}}$ by the mapping

$$C_n : C_{y_0}([0,T];N^+) \to C_{y_0}([0,T];D_n^+)$$

given by $C_n(y.)_t = c(y_{t \wedge \tau_n})$ where $c : \bar{D}_n \to D_n^+$ is the continuous map of the closure $\bar{D}_n$ to D_n^+ which sends the boundary of D_n to the point at infinity, leaving the other points unchanged. Moreover if $f : D_n \to \mathbf{R}$ is bounded and measurable with compact support in D_n, then

$$P_t^{\mathcal{B},n} f(y_0) = \mathbf{E}_{y_0}^{\mathcal{B}} \left[f(y_t) \chi_{\{t < \tau_n\}} \right] \qquad \text{for all } y_0 \in D_n. \tag{9.1}$$

K.D. Elworthy et al., *The Geometry of Filtering*, Frontiers in Mathematics, DOI 10.1007/978-3-0346-0176-4_9, © Springer Basel AG 2010

Using the notation of Chapter 4, let b be a vector field on N for which there is a T^*N-valued process α in $L^2_{\mathcal{B},\mathrm{loc}}$ such that

$$2\sigma^{\mathcal{B}}(\alpha_t) = b(y_t) \qquad 0 \leqslant t \leqslant T$$

for $\mathbf{P}_{y_0}$ almost all $y_. \in C_{y_0}([0,T]; N^+)$. Set

$$Z_t = \exp\{M^{\alpha}_t - \frac{1}{2}\langle M^{\alpha}\rangle_t\} \qquad 0 \leqslant t \leqslant T.$$

This exists by the non-explosion of the diffusion process generated by $\mathcal{B}$, and is a local martingale with $\mathbf{E}Z_t \leqslant 1$.

For bounded measurable $f : N \to \mathbf{R}$ define $Q_t f(y_0) = \mathbf{E}^{\mathcal{B}}_{y_0}[Z_t f(y_t)]$ for $y_0 \in N$. Since the pair $(y_., Z_.)$ is Markovian this determines a sub-Markovian semi-group on the space of bounded measurable functions. Also for bounded measurable $f : D_n \to \mathbf{R}$ define

$$Q^n_t f(y_0) = \mathbf{E}^{\mathcal{B}}_{y_0}[Z_{t \wedge \tau_n} f(y_t)\chi_{\{t<\tau_n\}}].$$

Lemma 9.1.2. *For any $T > 0$ the measures $\{\mathbf{Q}^n_{y_0}\}_{y_0 \in N}$ given by*

$$\mathbf{Q}^n_{y_0} = Z_{T \wedge \tau_n}\mathbf{P}^{\mathcal{B},n}_{y_0}$$

solve the martingale problem up to time T for $\mathcal{B}+b$ restricted to D_n. In particular, if the martingale problem for $\mathcal{B}+b$ on D_n has a unique solution, then for $f : D_n \to \mathbf{R}$ with compact support,

$$P^{\mathcal{B}+b,n}_t f(y_0) = Q^n_t f(y_0). \tag{9.2}$$

Proof. First note that $Z_.$ satisfies the usual stochastic equation which in our notation becomes:

$$Z_t = 1 + M^{Z\alpha}_t, \qquad 0 \leqslant t \leqslant T.$$

It becomes a martingale when stopped at any τ_n.

Let $f : N \to \mathbf{R}$ be C^2 with compact support. Use Itô's formula and the definition of M^{α} to see that

$$f(y_t)Z_t = f(y_0) + M^{Z(df)y.}_t + M^{fZ\alpha}_t + \int_0^t \mathcal{B}f(y_s)Z_s ds + \langle M^{df}, M^{Z\alpha}\rangle_t. \tag{9.3}$$

Now

$$\langle M^{df}, M^{Z\alpha}\rangle_t = 2\int_0^t df\left(\sigma^{\mathcal{B}}_{y_s}(Z_s\alpha_s)\right)ds \tag{9.4}$$

$$= \int_0^t df\left(Z_s b(y_s)\right)ds. \tag{9.5}$$

Thus

$$f(y_t)Z_t - f(y_0) - \int_0^t \mathcal{B}f(y_s)Z_s ds - \int_0^t df\left(Z_s b(y_s)\right)ds, \qquad 0 \le t \leqslant T,$$

is a local martingale under $\mathbf{P}_{y_0}^{\mathcal{B}}$. It is localised by the stopping times $\{\tau_n\}_n$.

Now suppose the support of f is in D_n. Let $\phi : C_{y_0}([0, T]; D_n^+) \to \mathbf{R}$ be $\mathcal{F}_r^{y_0}$-measurable and bounded, so $\phi \circ C_n$ will also be $\mathcal{F}_r^{y_0}$-measurable. Then, using Remark 9.1.1 above, and the martingale properties just mentioned together with Fubini's theorem, if $0 \leqslant r \leqslant t \leqslant T$,

$$\int \left[\left(f(z_t) - \int_0^t (\mathcal{B}+b)(f)(z_s)ds - f(z_r) - \int_0^r (\mathcal{B}+b)(f)(z_s)ds \right) \phi \right] Z_{T \wedge \tau_n} \ d\mathbf{P}_{y_0}^{\mathcal{B},n}(z.)$$

$$= \mathbf{E}_{y_0}^{\mathcal{B}} \left[\left(f(y_{t \wedge \tau_n}) - f(y_{r \wedge \tau_n}) - \int_{r \wedge \tau_n}^{t \wedge \tau_n} (\mathcal{B} + b)(f)(y_s)ds \right) Z_{T \wedge \tau_n} \phi \circ C_n \right]$$

$$= \mathbf{E}_{y_0}^{\mathcal{B}} \left[\left(f(y_{t \wedge \tau_n}) - f(y_{r \wedge \tau_n}) - \int_r^t (\mathcal{B} + b)(f)(y_s \wedge \tau_n)ds \right) Z_{T \wedge \tau_n} \phi \circ C_n \right]$$

$$= \mathbf{E}_{y_0}^{\mathcal{B}} \left[\left((f(y_{t \wedge \tau_n}) - f(y_{r \wedge \tau_n})) Z_{t \wedge \tau_n} - \int_{r \wedge \tau_n}^{t \wedge \tau_n} (\mathcal{B} + b)(f)(y_s)Z_s ds \right) \phi \circ C_n \right]$$

$$= 0$$

where the first integral is over the path space $C_{y_0}([0, T]; D_n^+)$. To obtain the second and third equalities we have used the fact that $(\mathcal{B} + b)f$ has compact support in D_n in order to remove and re-insert the stopping time τ_n: for example first remove and modify the upper limit of the range of integration by

$$\chi_{\{s<t \wedge \tau_n\}}(\mathcal{B} + b)f(y_s) = \chi_{\{s<t\}}\chi_{\{s<\tau_n\}}(\mathcal{B} + b)f(y_s) = \chi_{\{s<t\}}(\mathcal{B} + b)f(y_s \wedge \tau_n)$$

and then the lower limit by

$$\chi_{\{s<t\}}\chi_{\{s<\tau_n\}}\chi_{\{s>r \wedge \tau_n\}} = \chi_{\{s<t\}}\chi_{\{s<\tau_n\}}\chi_{\{s>r\}}.$$

Thus $\{\mathbf{Q}_{y_0}^n\}_{y_0 \in D_n}$ is a solution to the martingale problem. It follows that for such f, if we have uniqueness of the martingale problem, then

$$p_t^{\mathcal{B}+b}f(y_0) = \mathbf{E}^{\mathcal{B}} Z_{T \wedge \tau_n} f(y_t)\chi_{\{t<\tau_n\}}$$

$$= \mathbf{E}^{\mathcal{B}} Z_{t \wedge \tau_n} f(y_t)\chi_{\{t<\tau_n\}}$$

as required. □

In what follows in this section the assumption that b is locally Lipschitz is only used to ensure that uniqueness of the martingale problem holds for $\mathcal{B} + b$ together with its restrictions to each D_n, e.g. see [53], and could be replaced by assuming such uniqueness.

Theorem 9.1.3. *Suppose b is locally Lipschitz, Then:*

(i) *For all bounded measurable $f : N \to \mathbf{R}$,*

$$P_t^{\mathcal{B}+b} f(y_0) = \mathbf{E}_{y_0}^{\mathcal{B}} Z_t f(y_t) \chi_{\{t<\varsigma\}}. \tag{9.6}$$

(ii) *The probability that the diffusion process from y_0 generated by $\mathcal{B} + b$ has not exploded by time t is $\mathbf{E}_{y_0}^{\mathcal{B}} Z_t$.*

(iii) *If $\mathcal{B}+b$ is conservative, then $\{Z_t\}_t$ is a martingale under $\mathbf{P}_{y_0}^{\mathcal{B}}$ for all $y_0 \in N$.*

Proof. First suppose f has compact support. Then f has support in D_n for sufficiently large n. Trivially we have $Z_{t \wedge \tau_n} \chi_{\{t<\tau_n\}} = Z_t \chi_{\{t<\tau_n\}}$. From these facts and the dominated convergence theorem:

$$
\begin{aligned}
P_t^{\mathcal{B}+b} f(y_0) &= \mathbf{E}_{y_0}^{\mathcal{B}+b} f(y_t) \chi_{\{t<\varsigma\}} \\
&= \lim_n \mathbf{E}_{y_0}^{\mathcal{B}+b} f(y_t) \chi_{\{t<\tau_n\}} \\
&= \lim_n P_t^{\mathcal{B}+b,n} f(y_0) \\
&= \lim_n Q_t^n f(y_0) \\
&= \lim_n \mathbf{E}_{y_0}^{\mathcal{B}} [Z_{t \wedge \tau_n} f(y_t) \chi_{\{t<\tau_n\}}] \\
&= \lim_n \mathbf{E}_{y_0}^{\mathcal{B}} [Z_t f(y_t) \chi_{\{t<\tau_n\}}] \\
&= \mathbf{E}_{y_0}^{\mathcal{B}} [Z_t f(y_t) \chi_{\{t<\varsigma\}}].
\end{aligned}
$$

This can be extended to a general bounded measurable function f on N by writing $f = \lim_n \chi_{D_n} f$ to prove (i). In particular applying this when $f(x) = 1$ for all $x \in N$ we obtain (ii). But then (iii) is immediate. $\square$

From this we immediately obtain our version of the GMCM Theorem:

Theorem 9.1.4. *Suppose the diffusion operator $\mathcal{B}$ and its perturbation $\mathcal{B} + b$ by a locally Lipschitz vector field b on N are both conservative. Assume that $\mathcal{B} + b$ is cohesive or more generally that there is a locally bounded, measurable one-form $b^{\#}$ on N such that*

$$2\sigma_y^{\mathcal{B}}(b_y^{\#}) = b(y), \qquad y \in N.$$

Then

$$\exp\left(M_t^{b^{\#}} - \frac{1}{2} \langle M^{b^{\#}} \rangle_t\right), \qquad 0 \leqslant t \leqslant T$$

is a martingale under $\mathbf{P}^{\mathcal{B}}$ and for each $y_0 \in N$ the measures $\mathbf{P}_{y_0}^{\mathcal{B}}$ and $\mathbf{P}_{y_0}^{\mathcal{B}+b}$ on $C_{y_0}([0,T]; N)$ are equivalent with

$$\frac{d\mathbf{P}_{y_0}^{\mathcal{B}+b}}{d\mathbf{P}_{y_0}^{\mathcal{B}}} = \exp\left(M_T^{b^{\#}} - \frac{1}{2} \langle M^{b^{\#}} \rangle_T\right).$$

9.2 Stochastic differential equations for degenerate diffusions

Let $\mathcal{B}$ be a (smooth) diffusion operator on N. If its symbol $\sigma^{\mathcal{B}} : T^*N \to TN$ does not have constant rank there may be no smooth, or even C^2, factorisation

$$T^*N \xrightarrow{X^*} \underline{\mathbf{R}}^m \xrightarrow{X} TN$$

of $2\sigma_x^{\mathcal{B}}$ into $X(x)X^*(x)$ for $X : N \times \mathbf{R}^m \to TN$, as usual, for any finite dimensional m or even with $\mathbf{R}^m$ replaced by a separable Hilbert space. For completeness we first give the details of the constant rank case. Following that we describe a counterexample for the general situation. It is taken from Choi & Lam, [18], which also includes a brief history of the algebraic aspects of the problem. Then we look at variations of the problem and some positive, and other negative, results. However the discussion here is far from complete. For more concerning the expression of non-negative real-valued functions as sums of squares see Bony, Broglia, Colombini, & Pernazza, [13].

9.2.1 Square roots of symbols of constant rank

We have made much use of the following standard result:

Theorem 9.2.1. *Let $\pi : V \to M$ be a C^r vector bundle modelled on a Hilbert space G, for some $0 \leqslant r \leqslant \infty$, with a C^r vector bundle map $\Lambda : V^* \to V$. Suppose Λ satisfies the symmetry and positive semi-definiteness conditions:*

(i) $\Lambda^* = \Lambda$;

(ii) *if $\ell \in V^*$, then $\ell(\Lambda(\ell)) \geqslant 0$.*

If the image of Λ is a C^r subbundle of V, then there exists a Hilbert space H and a C^r vector bundle map $X : \underline{H} \to V$ of the trivial H-bundle to V such that $\Lambda = X^*X$. If the fibre space G is finite dimensional and Λ has constant rank, then the image of Λ is automatically such a subbundle and if also M is finite dimensional (and paracompact) we can take H to be finite dimensional.

Proof. Denote the image of Λ by W with inclusion $i : W \to V$. The basic idea of the proof is to show that Λ induces a Riemannian metric on W. Then embed W in a trivial bundle $\underline{H}^1$, by some $c : W \to \underline{H}^1$, extend the Riemannian metric over $\underline{H}^1$ and then take a metric preserving isomorphism $\Psi : \underline{H}^1 \to \underline{H}$ where $\underline{H}$ denotes the trivial H-bundle over M with trivial Riemannian structure. The map X is then the adjoint $\Psi \circ c$ composed with the inclusion i. The details follow:

Let $Z = \ker \Lambda$. Since Λ maps onto W it follows that Z is a C^r subbundle of V^*. Note that Z is the annihilator $W^\perp$ of W because of the symmetry of Λ. Moreover $i^* : V^* \to W^*$ is surjective with kernel $W^\perp$ so gives a vector bundle isomorphism $\alpha : V^*/Z \to W^*$. This way we obtain an isomorphism $\bar{\Lambda} : W^* \to W$

as the composition of α^{-1} with the map $\Lambda^0 : V^*/Z \to W$ obtained from Λ by quotienting out its kernel.

We claim that $\bar{\Lambda}$ is symmetric and positive semi-definite. To see this take a section $s : V^*/Z \to V^*$ of the projection (for example by giving V^* a Riemannian structure). Then $\bar{\Lambda} = \Lambda^0 \circ \alpha^{-1}$ with $\alpha = i^* \circ s$ and $i \circ \Lambda^0 = \Lambda \circ s$. Also if $\eta \in W^*$ set $\eta^0 = s(\alpha^{-1}(\eta)) \in V^*$, so that $\eta = i^*(\eta^0) = \eta^0 \circ i$. Thus if also $\xi \in W^*$ we have

$$\eta(\bar{\Lambda}\xi) = \eta^0 \circ i(\Lambda^0 \circ \alpha^{-1}(\xi))$$
$$= \eta^0(\Lambda(\xi^0)),$$

which is symmetric in ξ and η by the symmetry of Λ. This also shows the required positivity by taking $\xi = \eta$.

In fact $\bar{\Lambda}$ is an isomorphism and so is strictly positive. It therefore induces a Riemannian structure on W for which it is the canonical isomorphism, given by the Riesz transform. For suitable H^1 we can then find a C^r embedding $c : W \to \underline{H}$ onto a subbundle, see for example [61]. Now extend the Riemannian structure of $c[W]$ to one on the whole of $\underline{H}^1$. This determines a C^r map $A : M \to Pos(H^1)$, where $Pos(H^1)$ is the space of positive definite symmetric linear automorphisms of H^1, such that the metric over $x \in M$ is given in terms of the inner product $\langle -, - \rangle$ of H^1 by

$$\langle u, v \rangle_x = \langle A(x)u, v \rangle.$$

The map $S \mapsto \sqrt{S}$ from $Pos(H^1) \to Pos(H^1)$ is C^∞, for example see [61]. Let H be a copy of H^1 and define $\Psi : \underline{H}^1 \to \underline{H}$ by $\Psi(x, h) = (x, \sqrt{A(x)}h)$. Then $\langle \Psi(x, h), \Psi(x, k) \rangle = \langle A(x)h, k \rangle_x$ for $h, k \in H^1$ and $x \in M$. Thus $\Psi \circ c : W \to \underline{H}$ is isometric into the product Riemannian structure. Let $\bar{X} = (\Psi \circ c)^* : \underline{H} \to W$, an orthogonal projection. Then $\bar{\Lambda} = \bar{X}\bar{X}^*$. Now set $X = i \circ \bar{X} : \underline{H} \to V$. A similar calculation to that done to check the symmetry above shows that $XX^* = \Lambda$.

The claims about the case of finite dimensional G follow from results in [61] or other standard texts which cover vector bundle theory. $\square$

9.2.2   A smooth diffusion operator with no smooth Hörmander form

Let Q be the quartic polynomial in four variables given by

$$Q(x, y, z, w) = (w^4 + z^2x^2 + x^2y^2 + y^2z^2) - 4xyzw.$$

By the inequality of the geometric and arithmetic mean we see that Q takes only non-negative values so that

$$\mathcal{A}(f)(x, y, z, w) := Q(x, y, z, w)\frac{\partial^2 f}{\partial x^2} \tag{9.7}$$

defines a diffusion operator on $\mathbf{R}^4$ of non-constant rank. We will show this has no Hörmander form consisting of a sum of squares of twice differentiable vector fields possibly with a drift .

Lemma 9.2.2. *Suppose P is a polynomial in x, y, z, w satisfying $P(x, y, z, w)^2 \leqslant Q(x, y, z, w)$ for all real x, y, z, w. Then $P(0, 0, 0, w) = 0$ for all $w \in \mathbf{R}$.*

Proof [18]. Clearly P must be quadratic, with no linear or constant terms, and cannot contain x^2, y^2 or z^2. Following from this we see it cannot contain any of xw, yw, zw. This leaves only zx, xy, yz and w^2. Consideration of the order of the terms for small x, y, z finishes the proof. □

Now suppose that $Q(x, y, z, w) = \sum_{j=1}^{m}(f^j(x, y, z, w))^2$ where the functions f^j are twice differentiable at the origin. We easily see that each f^j vanishes together with its first derivative at the origin. It follows that each $(f^j)^2$ is 4-times differentiable at the origin with $D^4(f^j)^2(0)(h, h, h, h) = (D^2 f^j(0)(h, h))^2$ for $h \in \mathbf{R}^4$. Observe that $D^4 Q(0)(h, h, h, h) = 4!Q(h)$ and consider the Hessian of f^j, a quadratic form, evaluated at h as a polynomial in the components of h. Hence $(D^2 f^j(0)(h, h))^2 = D^4(f^j)^2(0)(h, h, h, h) \leqslant D^4 Q(0)(h, h, h, h) = 4!Q(h)$. The Lemma then implies that each $D^2 f^j(0)(h, h)$ vanishes when h is on the w-axis contradicting the fact that this is not true for $Q(h)$. This proves our non-existence assertion.

Note that the same proof shows that we cannot write Q as an infinite sum $Q(x, y, z, w) = \sum_{j=1}^{\infty}(f^j(x, y, z, w))^2$ of squares of functions which converges point-wise on $\mathbf{R}^4$ and which are twice differentiable at the origin with second derivative at the origin also converging. Moreover:

Proposition 9.2.3. *The symbol of $\mathcal{A}$ given by formula (9.7) has no factorisation into $X(x)X^*(x)$ for $X : \mathbf{R}^4 \times H \to \mathbf{R}^4$ which is twice differentiable at $(0, 0)$, for any separable Hilbert space H.*

Proof. If such a factorisation existed it would lead to

$$X(x, y, z, w)X^*(x, y, z, w)((1, 0, 0, 0)) = (Q(x, y, z, w), 0, 0, 0)$$

$$\text{for} \quad (x, y, z, w) \in \mathbf{R}^4.$$

Take an orthonormal base $\{e_j\}_j$ for H. Set $X^j(x, y, z, w) = X(x, y, z, w)(e_j) \in \mathbf{R}^4$. Let f_j be the first component of X^j. Then

$$\sum_j (f_j(x, y, z, w))^2 = Q(x, y, z, w) \quad \text{for} \quad (x, y, z, w) \in \mathbf{R}^4.$$

Also since $p \mapsto X(p) : \mathbf{R}^4 \to H^*$ is twice differentiable at the origin we see that the second derivatives of the partial sums at the origin, which is $Q(h, h)$, also converge. However we have just observed that that cannot happen. □

The proof of the proposition also follows from the result in the next section.

9.2.3   Non-existence of C^2 flow-like couplings

Of particular interest is the question of the existence of a, possibly locally defined, continuous or differentiable, stochastic flow whose one-point motions have as generator a given diffusion operator $\mathcal{A}$. We do not know of any definitative treatment of this question. The existence of such a local flow would follow from the existence of a sufficiently smooth Hörmander form for $\mathcal{A}$. For example a C^2 Stratonovich stochastic differential equation has a C^1 local flow. On the other hand the generator $\mathcal{A}^{(2)}$ of the two point motion of such a flow ξ gives a coupling of $\mathcal{A}$ with itself in the sense of Example 2.1.7. It is given by

$$\mathcal{A}^{(2)}(f \otimes g)(x, y) = \mathcal{A}(f)(x) + \mathcal{A}(g)(y) + \Gamma^{\xi}((df)_x, (dg)_y)$$

where $\Gamma^{\xi} : T^*M \times T^*M \to \mathbf{R}$ has

$$\Gamma^{\xi}((df)_x, (dg)_y) = \lim_{t \to 0} \frac{1}{t} \left(f(\xi_t(x) - f(x)) \right) \left(g(\xi_t(y)) - g(y) \right)$$

and, [6], [7], [66], is essentially the reproducing kernel k which appears in Section 8.2. Indeed

$$\Gamma^{\xi}((df)_x, (dg)_y) = (dg)_y \left(k(x, y)(df)_x \right).$$

For the coupling coming from such a flow the generator agrees with $\mathcal{A}$ when restricted to the diagonal, $\mathcal{A}^{(2)}(f \otimes g)(x, x) = \mathcal{A}(fg)(x)$, so that on the diagonal $\frac{1}{2}\Gamma^{\xi}$ agrees with the symbol of $\mathcal{A}$:

$$\Gamma^{\xi}\left((df)_x, (dg)_x \right) = 2(df)_x \sigma^{\mathcal{A}}((dg)_x) \qquad\qquad \text{for} \quad x \in M.$$

The existence of a smooth flow like coupling $\mathcal{A}^{(2)}$ therefore corresponds to the question of smooth *extendability of symbols*: given a smooth positive semi-definite bilinear $\sigma : T^M \oplus T^*M \to \mathbf{R}$ is there a C^r bilinear $\Gamma : T^*M \times T^*M \to \mathbf{R}$ which agrees with 2σ on the diagonal, is symmetric, and satisfies

$$\Gamma(u, v)^2 \leqslant 4\sigma(u, u)\sigma(v, v), \qquad u, v \in T^*M. \tag{9.8}$$

Note that this is weaker than the full positivity needed to have a reproducing kernel in order to obtain a flow.

Choi & Lam's example described above show that this is not in general possible for extensions for which the mixed second derivatives $D_2^2 D_1^2 \Gamma$ exist on the diagonal. To see this consider $\mathcal{A}$ given by equation (9.7). To show its symbol is not smoothly extendable, suppose $\gamma : \mathbf{R}^4 \times \mathbf{R}^4 \to \mathbf{R}$ satisfies $\gamma(a, a) = Q(a)$ and $\gamma(a, b)^2 \leqslant Q(a)Q(b)$ for all a, b in $\mathbf{R}^4$. Set $b = (0, 0, 0, w)$ for some $w \neq 0$. Then

$$\left(\frac{\gamma(a, b)}{\sqrt{Q(b)}} \right)^2 \leqslant Q(a)$$

and we see that $\gamma(0, b)$ and the partial derivative $D_1\gamma(0, b)$ both vanish and moreover the second partial derivative must satisfy $(D_1^2\gamma(0, b))^2 \leqslant Q(b)Q(-)$. Lemma

9.2.2 then implies that $D_1^2\gamma(0,b)((0,0,0,w'),(0,0,0,w')) = 0$ for all $w' \in \mathbf{R}$. Differentiation with respect to w then shows that the quartic determined by $D_2^2D_1^2(0,0)$ vanishes on the w-axis. Thus taking $a = (0,0,0,1)$ we have $Q(a) = 1$ and so

$$Q(a) = \frac{1}{4!}\frac{d^4}{dt^4}Q(ta)|_{t=0} = \frac{1}{4!}\frac{d^4}{dt^4}\gamma(ta,ta)|_{t=0} = \frac{1}{4!}D_1^2D_2^2\gamma(0,0)((a,a),(a,a))$$

which vanishes by the argument above, contradicting the fact that $Q(a) = 1$.

9.2.4 Locally Lipschitz square roots and Itô equations

The following is well known:

Theorem 9.2.4. *Let* $\sigma : \mathbf{R}^d \to \mathcal{L}_+(\mathbf{R}^m; \mathbf{R}^m)$ *be a* C^2 *map into the symmetric positive semi-definite* $(m \times m)$-*matrices, then* $\sqrt{\sigma} : \mathbf{R}^d \to \mathcal{L}_+(\mathbf{R}^m; \mathbf{R}^m)$ *is locally Lipschitz* .

For a proof see Freidlin [43], page 97 in [96] or Ikeda-Watanabe [53].

Corollary 9.2.5. *For a* C^2 *diffusion operator* $\mathcal{B}$ *on* N *there is a locally Lipschitz* $X : \underline{R}^m \to TN$ *with* $\sigma^{\mathcal{B}} = XX^*$ *for some* m.

Proof. For some m, take a smooth inclusion $TN \overset{i}{\to} \mathbf{R}^m$ as a subbundle (*e.g.* by embedding N in $\mathbf{R}^m$) and extend $\sigma^{\mathcal{B}}$ trivially to $\sigma_x^{\mathcal{B}} : N \to \mathcal{L}((\mathbf{R}^m)^*; \mathbf{R}^m)$ by

$$(\mathbf{R}^m)^* \overset{i_x^*}{\to} T_x^*N \overset{\sigma_x^{\mathcal{B}}}{\to} T_xN \overset{i_x}{\to} \mathbf{R}^m$$

identifying $(\mathbf{R}^m)^*$ with $\mathbf{R}^m$ and take the square root. $\square$

Let ∇ be a connection on a subbundle G of TN and let $X : \underline{R}^m \to G$ be a locally Lipschitz bundle map. Let A be a locally Lipschitz vector field on N. As in Elworthy [30] (p184) we can form the Itô stochastic differential equation on N,

$$(\nabla) \qquad dx_t = X(x_t)dB_t + A(x_t)dt$$

where (B_t) is a Brownian motion on $\mathbf{R}^m$. For given $x_0 \in N$ there will be a unique maximal solution $\{x_t : 0 \leqslant t < \zeta^{x_0}\}$ as usual, where by a solution we mean a sample continuous adapted process such that, for all C^2 functions $f : N \to \mathbf{R}$,

$$f(x_t) = f(x_0) + \int_0^t (df)_{x_s}X(x_s)dB_s + \int_0^t (df)_{x_s}A(x_s)ds$$

$$= \int_0^t \sum_{j=1}^m \nabla_{X^j(x_s)}(df|_G)X^j(x_s)ds.$$

Indeed in a local coordinate (U,ϕ) system the equation is represented by

$$dx_t^\phi = X_\phi(x_t^\phi)dB_t - \frac{1}{2}\sum_{j=1}^m \Gamma_\phi(x_t^\phi)\left(X_\phi^j(x_t^\phi)\right)\left(X_\phi^j(x_t^\phi)\right)dt + A_\phi(x_t^\phi)dt,$$

where X_ϕ, X^i_ϕ, and A_ϕ are the local representations of X, X^i and A, and Γ_ϕ is the Christoffel symbol.

Note that the generator of the solution process has symbol $\sigma_x = X(x)X(x)^*$, $x \in N$, and so a Lipschitz factorisation of $\sigma^{\mathcal{B}}$ together with a suitable choice of A will give a diffusion process with generator $\mathcal{B}$.

If in addition we have another generator G on N given in Hörmander form

$$G = \sum_{k=1}^{p} \mathbf{L}_{Y^k}\mathbf{L}_{Y^k} + \mathbf{L}_{Y^0}$$

for $Y^0, Y^1, \ldots, Y^k$ vector fields of class C^2 we can consider an SDE of mixed type

$$(\nabla) \qquad dx_t = \sum_{k=1}^{p} Y^k(x_t) \circ d\tilde{B}^k_t + X(x_t)dB_t + (Y^0(x_t) + A(x_t))dt$$

for $\tilde{B}^1, \ldots, \tilde{B}^k$ independent Brownian motions on $\mathbf{R}$ independent of (B_t). For a C^2 map $f : N \to \mathbf{R}$, a solution $\{x_t : 0 \leqslant t < \zeta^{x_0}\}$ will satisfy

$$f(x_t) = f(x_0) + \int_0^t (df)_{x_s} X(x_s)dB_s + \int_0^t \sum_{k=1}^{n} (df)_{x_s}(X^k(x_s))d\tilde{B}^k_s$$

$$= \int_0^t (\mathcal{B} + G)f(x_s)ds, \qquad t < \zeta^{x_0},$$

giving the unique solution to the martingale problem for $\mathcal{B}+G$. These SDE's fit into the general frame work of the 'Itô bundle' approach of Belopolskaya-Dalecky [8], see [47] and the Appendix of Brzezniak-Elworthy [15]; also see Emery [40](section 6.33, page 85) for a more semi-martingale oriented approach.

9.2.5   Miscellaneous results

A.   A factorisation with $X : N \times H \to TN$, for H a separable Hilbert space, can be found following Stroock and Varadhan, Appendix in [97], with the property that X is continuous and each vector field X^j is C^∞, where $X^j(x) = X(x)(e^j)$ for an orthonormal basis $(e_j)_{j=1}^\infty$ of H. However it seems unclear if such an X can be found with each $x \mapsto X(x)e$, $e \in H$, smooth.

B.   A non-negative $C^{3,1}$-function can be written as a finite sum of squares of $C^{1,1}$-functions by results of Fefferman & Phong and Guan, see [100]. Thus our operator $\mathcal{A}$ given by equation (9.7) is the generator of the solution of an SDE on $\mathbf{R}^4$ of the form $dx_t = \sum_{j=1}^{m} \lambda^j(x_t)e_1 \, dB^j_t$ where the λ^j are C^1 with locally Lipschitz derivatives, the B^j are independent one-dimensional Brownian motions, and $e_1 = (1,0,0,0)$. Equivalently we have a Stratonovich SDE $dx_t = \sum_{j=1}^{m} \lambda^j(x_t)e_1 \circ dB^j_t + A(x_t) \, dt$ where A is locally Lipschitz. See also Bony et al., [13]. It seems that this does not extend to more general $\mathcal{A}$, see [86], [14].

9.3 Semi-martingales and Γ-martingales along a Subbundle

Several of the concepts we have defined for diffusions also have versions for semi-martingales, and these are relevant to the discussion of non-Markovian observations in Chapter 5. Only *continuous* semi-martingales will be considered. Let S denote a subbundle of the tangent bundle TM to a smooth manifold M.

Definition 9.3.1. A semi-martingale $y_s, 0 \leqslant s < \tau$ is said to be *along S* if whenever ϕ is a C^2 one-form on M which annihilates S we have vanishing of the Stratonovich integral of ϕ along $y.$:

$$\int_0^t \phi_{y_s} \circ dy_s = 0 \quad 0 < t < \tau.$$

For simplicity take y_0 to be a point of M.

Proposition 9.3.2. *The following are equivalent:*

1. *the semi-martingale $y.$ is along S;*

2. *if $\alpha_s : 0 \leqslant s < \tau$ is a semi-martingale with values in the annihilator of S in T^*M, lying over $y.$, then*

$$\int_0^t \alpha_{y_s} \circ dy_s = 0 \quad 0 < t < \tau;$$

3. *for some, and hence any, connection Γ on S the process $y.$ is the stochastic development of a semi-martingale $y_s^\Gamma, 0 \leqslant s < \tau$ on the fibre S_{y_0} of S above y_0.*

If $\mathcal{L}$ is a diffusion operator, then the associated diffusion processes are all along S if and only if $\mathcal{L}$ is along S in the sense of Section 1.3.

Proof. Let $//.$ denote the parallel translation along the paths of $y.$ using Γ. If (3) holds, then

$$dy. = //. \circ dy^\Gamma.$$

and it is immediate that (2) is true. Also (2) trivially implies (1).

Now suppose that (1) holds. Let Γ be a connection on E and Γ^0 some extension of it to a connection on TM, so that the corresponding parallel translation $//^0$ will preserve S and some complementary subbundle of TM. Let y^{Γ^0} be the stochastic anti-development of $y.$ using this connection. To show that (3) holds it suffices to show that y^{Γ^0} takes values in S_{y_0}. For this choose a smooth vector bundle map $\Phi : TM \to M \times \mathbf{R}^m$ whose kernel is precisely S and let $\phi : TM \to \mathbf{R}^m$ denote its principal part and $\phi^j, j = 1, \ldots, m$ the components of ϕ. These are one-forms, which annihilates S. Then, for each j,

$$0 = \int_0^t \phi_s \circ dy_s = \int_0^t \phi_s //_s^0 \circ dy_s^{\Gamma^0} \quad 0 < t < \tau.$$

By the lemma below we see that $y_s^{\Gamma^0} \in S_{y_0}$ for each s, almost surely, and the result follows .

Finally suppose that y. is a diffusion process with generator $\mathcal{L}$. By Lemma 4.1.2 we have

$$M_t^\alpha = \int_0^t \alpha_{y_s} \circ dy_s - \int_0^t (\delta^{\mathcal{L}}\alpha)(y_s)ds, \qquad 0 \leqslant t < \zeta. \qquad (9.9)$$

for any C^2 one-form α. Suppose α annihilates S. Then if y. is along S both the martingale and finite variation parts of $\int_0^{\cdot} \alpha_{y_s} \circ dy_s$ vanish and so $(\delta^{\mathcal{L}}\alpha)(y_s) = 0$ almost surely for almost all $0 \leqslant s < \tau$. If this is true for all starting points we see $\mathcal{L}$ is along S. On the other hand if $\mathcal{L}$ is along S and α annihilates S we see that M^α vanishes by its characterisation in Proposition 4.1.1, since $\sigma^{\mathcal{L}}$ takes values in S. Thus both the martingale and finite variation parts of $\int_0^{\cdot} \alpha_{y_s} \circ dy_s$ vanish, and so the integral itself vanishes and the diffusion processes are along S. $\square$

Lemma 9.3.3. *Suppose z. and Λ. are semi-martingales with values in a finite dimensional vector space V and the space of linear maps $\mathcal{L}(V; W)$ of V into a finite dimensional vector space W, respectively. Let V_0 denote the kernel of Λ_s which is assumed non-random and independent of $s \geqslant 0$. Assume*

$$\int_0^{\cdot} \Lambda_s \circ dz_s = 0.$$

Then z. lies in V_0 almost surely.

Proof. We can quotient out by V_0 to assume that $V_0 = 0$, so we need to show that z. vanishes. Giving W an inner product, let $P_s : W \to \Lambda_s[V]$ be the orthogonal projection. Compose this with the inverse of Λ_s considered as taking values in $\Lambda_s[V]$, to obtain an $\mathcal{L}(W; V)$-valued semi-martingale $\tilde{\Lambda}$. formed by left inverses of Λ.. By the composition law for Stratonovich integrals

$$z_t = \int_0^t dz_s = \int_0^t \tilde{\Lambda}_s \Lambda_s \circ dz_s = \int_0^t \tilde{\Lambda}_s \circ d\Big(\int_0^s \Lambda_r \circ dz_r \Big) = 0 \qquad (9.10)$$

as required. $\square$

Let Γ be a connection on S. Note that by the previous proposition any semi-martingale y. which is along S has a well-defined anti-development y^Γ, say , which is a semi-martingale in S_{y_0}.

Definition 9.3.4. An M-valued semi-martingale is said to be a Γ-martingale if its anti-development using Γ is a local martingale.

Also we can make the following definition of an Itô integral of a differential form, using the analogue of a characterisation by Darling, [23], for the case $S = TM$;

Definition 9.3.5. If α is a predictable process with values in T^*M, lying over our semi-martingale y, define its Itô integral, $(\Gamma) \int_0^t \alpha_s dy_s$ along the paths of y with respect to Γ by

$$(\Gamma) \int_0^t \alpha_s dy_s = \int_0^t \alpha_s /\!/_s dy^\Gamma \qquad (9.11)$$

whenever the (standard) Itô integral on the right-hand side exists.

As usual this Itô integral is a local martingale for all suitable integrands α if and only if the process y is a Γ-martingale.

9.4 Second fundamental forms and shape operators

For a detailed treatment of the differential geometry of submanifolds of Riemannian manifolds see Chapter VII of Kobayashi &Nomizu Volume II, [56]. Here we just recall some basic formulae.

Let p be a point of a submanifold P of a Riemannian manifold Q. Let ∇^Q and ∇^P denote the Levi-Civita connections of Q and P respectively. We consider TP as a subbundle of $TQ|_P$ and let $TP^\perp$ denote the normal bundle. Suppose U and V are vector fields on P and Z is a smooth section of the normal bundle. These can all be extended smoothly to vector fields on Q and we let $\nabla^Q_{V(p)}U$ and $\nabla^Q_{V(p)}Z$ denote the covariant derivatives of these extensions in the direction $V(p)$. It is easy to see that they do not depend on the extensions.

Define $\alpha_p(V(p), U(p))$ to be the normal component of $\nabla^Q_{V(p)}U$. It depends only on the values of U and V at p, and gives a symmetric bilinear map $\alpha : TP \oplus TP \to TP^\perp$, This is the *second fundamental form* of P.

Define $A_{Z(p)}(V(p))$ to be the negative of the tangential component of $\nabla^Q_{V(p)}Z$. It depends only on the values of Z and V at p and gives a bilinear mapping $A : TP \oplus TP^\perp \to TP$. This is the *shape operator* of P at p. It satisfies:

$$\langle A_z(v), u \rangle_p = \langle \alpha_p(v, u), z \rangle_p, \qquad u, v \in T_pP \quad z \in T_pP^\perp. \qquad (9.12)$$

Gauss's formula is

$$\nabla^Q_{V(p)}U = \alpha_p(V(p), U(p)) + \nabla^P_{V(p)}U, \qquad (9.13)$$

and Weingarten's is

$$\nabla^Q_{V(p)}Z = -A_{Z(p)}(V(p)) + \nabla^\perp_{V(p)}Z, \qquad (9.14)$$

where $\nabla^\perp$ refers to covariant differentiation using the induced connection on the normal bundle.

Since these are local equations they apply equally well to manifolds P isometrically immersed in Q.

Finally recall that P is *minimal* if and only if the second fundamental form has zero trace at each point, that is if $\sum \alpha_p(e_i, e_i) = 0$ using an orthonormal basis of T_pP, for each p. It is said to be *totally geodesic* when a geodesic starting at any point of P in a direction tangential to P does not leave P for a positive amount of time (and so never leaves P if P is closed). This holds if and only if the second fundamental form, or equivalently the shape operator, vanishes identically, see [56] page 59.

9.5 Intertwined stochastic flows

In this section we shall consider the situation of a stochastic flow lying above another, first in relation to the corresponding properties of their reproducing kernels and reproducing kernel Hilbert spaces. As a by-product we will obtain a decomposition of the generator of the one-point motion of the flow "upstairs" which may not agree with the canonical one obtained in Chapter 2. As usual $p : N \to M$ will denote a smooth map, which we shall assume to be surjective. For simplicity we treat only smooth, i.e. C^∞, flows, with correspondingly smooth reproducing kernels. *We are not making any constant rank hypothesis.*

9.5.1 Intertwined reproducing kernels and Gaussian spaces of vector fields

Recall from Chapter 8 that associated to a stochastic flow $\tilde{\xi}$, say, on N we have a reproducing kernel $\tilde{k}$ with $\tilde{k}(u, v) : T_u N^* \to T_v N$ linear for each $u, v \in N$. In fact it gives maps from $\tilde{E}_u^*$ to $\tilde{E}_v$, where $\tilde{E}_u$ is the image of the one-point generator, $\mathcal{B}$ say, but we will not need that extra refinement here. The kernel can be obtained from the generator $\mathcal{B}^{(2)}$ on $N \times N$ of the two-point motion, see the discussion in Section 9.2.3. Indeed, from that discussion, identifying $T^*_{(u,v)}(N \times N)$ with $T_u^* N \times T_v^* N$, we see that the symbol of $\mathcal{B}^{(2)}$ is given by

$$(a^1, a^2)\sigma^{\mathcal{B}^{(2)}}(b^1, b^2) = a^1\sigma^{\mathcal{B}}(b^1) + a^2\sigma^{\mathcal{B}}(b^2) + \frac{1}{2}\left(a^2(\tilde{k}(u,v)b^1) + b^2(\tilde{k}(u,v)a^1)\right) \tag{9.15}$$

for $a^1, b^1 \in T_u^* N$ and $a^2, b^2 \in T_v^* N$.

Associated to such a reproducing kernel, $\tilde{k}$, there is its reproducing kernel Hilbert space of smooth vector fields, $\tilde{\mathcal{H}}$ say, so that $\tilde{k}(u, \cdot)(-) = \tilde{\rho}_u^* : T_u^* N \to \tilde{\mathcal{H}}$ where $\tilde{\rho}_u : \tilde{\mathcal{H}} \to T_u N$ is the evaluation. There is also the Gaussian family of vector fields, to be denoted by $\tilde{W}(\cdot)$, uniquely determined by

$$b\tilde{k}(u,v)(a) = \mathbf{E}\left[a(\tilde{W}(u))\, b(\tilde{W}(v))\right], \qquad a \in T_u^* N,\ b \in T_v^* N. \tag{9.16}$$

Note that for a, b as above,

$$\langle \tilde{k}(u, \cdot)(a), \tilde{k}(v, \cdot)(b)\rangle_{\tilde{\mathcal{H}}} = \langle \tilde{\rho}_u^*(a), \tilde{\rho}_v^*(b)\rangle_{\tilde{\mathcal{H}}} = b\tilde{k}(u,v)(a). \tag{9.17}$$

Proposition 9.5.1. *The following conditions on $\bar{k}$ are equivalent:*

(i) *The symbol of the two-point motion of the associated flow has $(p \times p)$-projectible symbol.*

(ii) *For any $x \in M$ if $p(u) = p(v) = x$, then, as elements of $\tilde{\mathcal{H}}$,*

$$\tilde{k}(u, \cdot)(T_u^* p(\ell)) = \tilde{k}(v, \cdot)(T_v^* p(\ell)) \qquad \text{for any } \ell \in T_x^* M.$$

(iii) *In the notation of* (ii),

$$T_u^*(\ell)\tilde{k}(u, u)(T_u^* p(\ell)) + T_v^*(\ell)\tilde{k}(v, v)(T_v^* p(\ell)) - 2T_v^*(\ell)\tilde{k}(u, v)(T_u^* p(\ell)) = 0.$$

(iv) *If $p(u) = p(v)$, then $T_u p\tilde{W}(u) = T_v p\tilde{W}(v)$ almost surely.*

Proof. First note that (ii) and (iii) are equivalent by expanding

$$\|\tilde{k}(u, \cdot)(T_u^* p(\ell)) - \tilde{k}(v, \cdot)(T_v^* p(\ell))\|_{\tilde{\mathcal{H}}}^2$$

and applying equation (9.17). To bring in (iv) write it as

$$\mathbf{E}[\left(\ell(T_u p\tilde{W}(u) - T_v p\tilde{W}(v))\right)^2] = 0 \qquad \text{for all } \ell \in T_x M,$$

then expand and use the relation (9.16) to see that it is equivalent to (iii). The equivalence of (i) with (ii) is immediate from equation (9.15) and Lemma 2.1.1. $\square$

Assuming criterion (ii) above we see that we have a unique family of linear maps $k(x, y) : T_x^* M \to T_y M$ for x and y in M such that

$$T_v p\, \tilde{k}(u, v)T_u^* p = k(p(u), p(v)) \qquad \text{for all } u, v \in N. \tag{9.18}$$

It is clear that k inherits the positivity conditions from $\tilde{k}$ that are required of a reproducing kernel: namely

(a) $k(y, x)^* = k(x, y)$,

(b) if $x_j, j = 1, \ldots, k$ are in M and $\ell_j \in T_{x_j} M$, then

$$\sum_{i,j=1}^{k} \ell_j k(x_i, x_j)\ell_i \geqslant 0.$$

If k is smooth we will say that the reproducing kernel $\tilde{k}$, or equivalently $\tilde{\mathcal{H}}$, or $\tilde{W}$, is *projectible over p* , or *p-projectible*, and that it *lies over* k or is *p-related* to k. Smoothness of k is inherited from that of $\tilde{k}$ when p is a submersion, as is easily seen from the local product structure of submersions. It is equivalent to the smoothness of the elements of $\mathcal{H}$.

We shall say that a Hilbert space $\tilde{\mathcal{H}}$ of vector fields on N *lies over* or *is p-related to* a Hilbert space of vector fields $\mathcal{H}$ on M if each element $h \in \tilde{\mathcal{H}}$ is p-related to some element $p_*(h)$ of $\mathcal{H}$ and the map $p_* : \tilde{\mathcal{H}} \to \mathcal{H}$ is an orthogonal projection (i.e. p_* is an isometry on the orthogonal complement of its kernel, or equivalently its adjoint maps $\mathcal{H}$ isometrically to a subspace of $\tilde{\mathcal{H}}$).

Theorem 9.5.2. *Let $\tilde{k}$ and k be smooth reproducing kernels for vector fields on N and M respectively, with reproducing kernel Hilbert spaces $\tilde{\mathcal{H}}$ and $\mathcal{H}$. Then $\tilde{k}$ lies over k if and only if $\tilde{\mathcal{H}}$ lies over $\mathcal{H}$.*

Proof. First suppose $\tilde{k}$ lies over k. Recall that the set $\{k(x, \cdot)(\ell) : x \in M, \ell \in T_x^* M\}$ is total in $\mathcal{H}$. Define the "horizontal subspace" $\mathcal{H}^H$ of $\tilde{\mathcal{H}}$ to be the closed linear span of $\{\tilde{k}(u, \cdot)(T_u p^*(\ell)) : u \in N, \ell \in T_{p(u)}^* M\}$. Using the property (9.17) note that for $u_j \in N$ and $\ell_j \in T_{x_j}^* M$ for $j = 1, \ldots, k$ and $x_j = p(u_j)$ we have

$$\| \sum_{j=1}^{k} \tilde{k}(u_j, \cdot)(T_{u_j} p^*(\ell_j)) \|_{\tilde{\mathcal{H}}}^2 = \sum_{i,j=1}^{k} T_{u_j} p^*(\ell_j)\{\tilde{k}(u_i, u_j)(T_{u_i} p^*(\ell_i))$$

$$= \sum_{i,j=1}^{k} \ell_j k(x_i, x_j) \ell_i = \| \sum_{j} k(x_j, \cdot)(\ell_j) \|_{\mathcal{H}}^2 .$$

Since equation (9.18) tells us that $\tilde{k}(u, \cdot)(T_u p^*(\ell))$ is p-related to $k(p(u), \cdot)(\ell)$ for all relevant u and ℓ, we see that we have an isometry $h \mapsto p_*(h)$ of $\mathcal{H}^H$ onto $\mathcal{H}$ extending the map $\tilde{k}(u, \cdot)(T_u p^*(\ell)) \mapsto k(p(u), \cdot)(\ell)$, with h always p-related to $p_*(h)$.

Suppose now that $h \in \tilde{\mathcal{H}}$ is orthogonal to $\mathcal{H}^H$. Then we claim $T_u p(h(u)) = 0$ for all $u \in N$, i.e. h is vertical, so h is p-related to the zero vector field, and we set $p_*(h) = 0$. To check this claim set $p(u) = x$ and take $\ell \in T_x^* M$. Then

$$\ell (T_u p(h(u))) = T_u^* p(\ell)(\tilde{\rho}_u(h)) = \langle \tilde{\rho}_u^*(T_u^* p(\ell)), h \rangle_{\tilde{\mathcal{H}}} = \langle \tilde{k}(u, \cdot) T_u^*(\ell), h \rangle_{\tilde{\mathcal{H}}} = 0$$

as required.

For the converse assume that $\tilde{\mathcal{H}}$ lies over $\mathcal{H}$ with projection $p_* : \tilde{\mathcal{H}} \to \mathcal{H}$. If $\tilde{W}(\cdot)$ and $W(\cdot)$ are the Gaussian vector fields corresponding to $\tilde{\mathcal{H}}$ and $\mathcal{H}$ it follows that $p_* \tilde{W}(u)$ and $W(p(u))$ are equal in law for each $u \in N$. Now, for $\alpha \in T_{p(u)}^* M$ and $\beta \in T_{p(v)}^* M$, take $a = T_u^* p(\alpha)$ and $b = T_v p^*(\beta)$ in equation (9.16). This gives

$$\beta \left(T_v p \tilde{k}(u, v)(T_u^* p(\alpha)) \right) = T_v p^*(\beta) \tilde{k}(u, v)(T_u^* p(\alpha))$$

$$= \mathbf{E} \left[\alpha((T_u p(\tilde{W}(u))) \, \beta(T_v p(\tilde{W}(v))) \right]$$

$$= \mathbf{E} \left[\alpha(W(p(u)) \, \beta(W(p(v))) \right]$$

$$= \beta \left(k(p(u), p(v))(\alpha) \right)$$

so $\tilde{k}$ lies over k. $\qquad\square$

Remark 9.5.3. From the theorem, if $\tilde{k}$ lies over k, we have a decomposition of $\tilde{\mathcal{H}}$ into $\mathcal{H}^H \oplus (\mathcal{H}^H)^\perp$ where $(\mathcal{H}^H)^\perp$ is the kernel of p_* and from the proof we see that $(\mathcal{H}^H)^\perp$ consists only of vertical vector fields; indeed it must contain all the vertical vector fields in $\tilde{\mathcal{H}}$. On the other hand, elements of $\mathcal{H}^H$ may take vertical values at some points. As an example let $p : \mathbf{R}^2 \to \mathbf{R}$ be the projection onto the first co-ordinate. Let $\mathcal{H}$ be one-dimensional and generated by h for $h(x) = x^2$ assigned norm equal to 1. Let $\tilde{\mathcal{H}}$ have orthonormal base $\{h^1, h^2\}$ given by $h^1(x,y) = (x^2, \cos(y))$ and $h^2(x,y) = (0, \sin(y))$. Then $\mathcal{H}^H = \mathbf{R}h^1$. The situation is clarified in the next proposition.

Proposition 9.5.4. *Using the notation of the theorem suppose $\tilde{k}$ lies over k. Let H_u denote the horizontal subspace of $T_u N$ as defined after Proposition 2.1.2, (it was not necessary to assume that the symbol of $\mathcal{A}$ has constant rank to define H_u). Then for $h \in \mathcal{H}$ the "lift" $(p_*)^*(h) \in \mathcal{H}^H$ of h has $(p_*)^*(h)(u) \in H_u$ if $h \in (\ker \rho_{p(u)})^\perp$. If h is in the "redundant noise" subspace at $p(u)$, i.e. $h \in \ker \rho_{p(u)}$, then $(p_*)^*(h)(u)$ is vertical.*

Proof. Fix $x \in M$ and $u \in p^{-1}(x)$. Since in this section we are considering kernels such as k as giving maps $k(x,y) : T_x^* M \to T_y M$ it will be convenient to take a splitting of $T_x M$ into E_x, the image of $\rho_x : \mathcal{H} \to T_x M$, and a complementary subspace so that we can use the inner product on E_x induced from ρ_x to consider the mapping $v \mapsto v^\#$ as a map from E_x to $T_x^* M$.

Let $K^\perp(x) : \mathcal{H} \to \mathcal{H}$ be the projection onto the kernel of ρ_x. Then $K^\perp(x)(h) = (\rho_x)^*(\rho_x(h))^\# = k(x,\cdot)(h(x)^\#)$ and so h is in the orthogonal complement of the redundant noise at x if and only if $h(\cdot) = k(x,\cdot)(h(x)^\#)$. If so we know from the proof of Theorem 9.5.2 that the lift, $(p_*)^*(h)$ of h, is just $\tilde{k}(u,\cdot)(T_u p^*(h(x)^\#))$. In particular

$$(p_*)^*(h)(u) = \tilde{k}(u,u)(T_u p^*(h(x)^\#)). \qquad (9.19)$$

On the other hand we can apply the discussion in Section 2.3, especially equation (2.22), to the p-related SDE's $\tilde{X}$ and X given by $\tilde{X}(v) = \tilde{\rho}_v : \tilde{\mathcal{H}} \to T_v N$ and $X(y) = \rho_y \circ p_* : \tilde{\mathcal{H}} \to T_y M$ to see that the horizontal lift of $h(x)$ is given by

$$h_u(h(x)) = \tilde{\rho}_u(p_*)^* \rho_x^*(h(x)^\#) = \tilde{\rho}_u(p_*)^* k(x,\cdot)(h(x)^\#)$$
$$= \tilde{k}(u,u)(T_u p)^*(h(x)^\#)$$
$$= (p_*)^*(h)(u)$$

by equation (9.19), as required.

In the situation where h is in the redundant noise subspace at x, so $h(x) = 0$, take $\ell \in T_x^* M$ and observe that:

$$\ell(T_u p \, \tilde{\rho}_u((p_*)^*(h))) = \langle \tilde{k}(u,\cdot)(T_u p)^*(\ell), (p_*)^*(h) \rangle_{\tilde{\mathcal{H}}}$$
$$= \langle k(x,\cdot)(\ell), h \rangle_{\mathcal{H}}$$
$$= \ell(h(x)) = 0. \qquad \square$$

Remark 9.5.5. Theorem 9.5.2 shows that, if we have a diffusion operator $\mathcal{B}$ over a diffusion operator A, then if $\mathcal{B}$ has a Hörmander form representation

$$\mathcal{B} = \frac{1}{2} \sum_j \mathbf{L}_{\tilde{X}^j} \mathbf{L}_{\tilde{X}^j}$$

with $\tilde{X}^1, \tilde{X}^2, \ldots$ an orthonormal base for a possibly infinite dimensional Hilbert space $\tilde{\mathcal{H}}$ of smooth vector fields on N such that the corresponding reproducing kernel is projectible over p, then there is a decomposition $\mathcal{B} = A + \mathbb{B}^V$ of $\mathcal{B}$ into diffusion operators such that $\mathbb{B}^V$ is vertical and A lies over $\mathcal{A}$. Unlike the decomposition of cohesive operators given in Theorem 2.2.5 this is not canonical and depends on the choice of such a Hörmander form, even assuming such a projectible smooth Hörmander form exists. It agrees with the usual decomposition $\mathcal{B} = \mathcal{A}^H + \mathcal{B}^V$ if we have a smooth Hörmander form for $\mathcal{B}^V$ which we use together with the lift of such a form for $\mathcal{A}$ to produce the required one for $\mathcal{B}$.

Example 9.5.6. Take $M = \mathbf{R}$ and $N = \mathbf{R}^2$ with $p(x, y) = x$. Let $\tilde{\mathcal{H}}$ be three-dimensional with orthonormal base the vector fields $\tilde{h}^1, \tilde{h}^2, \tilde{h}^3$ given by

$$\tilde{h}^1(x,y) = (\sin x, y), \quad \tilde{h}^2(x,y) = (\cos x, 1), \quad \tilde{h}^3(x,y) = (0,1).$$

Then $\mathcal{H}$ has orthonormal base $\{h^1, h^2\}$ where $h^1(x) = \sin x$ and $h^2(x) = \cos x$. (So $\mathcal{A} = \frac{1}{2}\Delta$.) The only elements of $\tilde{\mathcal{H}}$ which only take vertical values are the scalar multiples of $\tilde{h}^3$. Therefore $\mathcal{H}^H$ is spanned by $\tilde{h}^1$ and $\tilde{h}^2$ and these are the "lifts" of h^1 and h^2 respectively. Note that at $(0,1)$ this lift of h^1 is vertical. Also, at (x,y),

$$A = \frac{1}{2}\frac{\partial^2}{\partial x^2} + \frac{1}{2}(1+y^2)\frac{\partial^2}{\partial y^2} + (\cos x + y \sin x)\frac{\partial^2}{\partial x \partial y}.$$

9.5.2   Intertwined stochastic flows that induce Levi-Civita connections

Consider a Riemannian submersion $p : N \to M$. Is it possible to construct a smooth stochastic flow $\tilde{\xi}.$ of Brownian motions on N that lies over a stochastic flow on M and such that the connection determined by $\tilde{\xi}.$, in the sense of Section 8.2, is the Levi-Civita connection? We shall show that this is so if and only if the submersion has totally geodesic fibres and the horizontal distribution on N determined by the Riemannian structures is integrable, at least when N is compact. This is essentially "local triviality" of the Riemannian structure of the submersion, see [83], and the discussion in the Notes on Chapter 7, Section 7.3.

   In general, given $p : N \to M$ we will say that a, possibly local, stochastic flow $\tilde{\xi}.$ on M *lies over* such a flow $\xi.$ on M, or that $\tilde{\xi}.$ and $\xi.$ are *intertwined* by p if:

(i) They are defined on the same probability space $\{\Omega, \mathcal{F}, \mathbf{P}\}$.

(ii) The lifetimes $\zeta : M \times \Omega \to (0, \infty]$ and $\tilde{\zeta} : N \times \Omega \to (0, \infty]$ of $\xi.$ and $\tilde{\xi}.$ almost surely satisfy

$$\tilde{\zeta}(y, \omega) \leqslant \zeta(p(y), \omega) \qquad \text{for all } y \in N.$$

(iii) Almost surely, for all $y \in N$ we have

$$p\left(\tilde{\xi}_t(y, \omega)\right) = \xi_t(p(y), \omega) \qquad t < \tilde{\zeta}(y, \omega).$$

Recall that a local stochastic flow on manifold M determines a reproducing kernel k and a *drift* vector field, A say, and up to law it is determined by them as the solution flow of the SDE on M:

$$dx_t = \rho_{x_t} \circ dW_t + A(x_t) \, dt, \tag{9.20}$$

for a Wiener process $\{W_t\}_{t \geqslant 0}$ of vector fields determined up to law by the reproducing kernel Hilbert space $\mathcal{H}$ where, as usual, ρ_x denotes the evaluation map, evaluated at the point x of M.

Proposition 9.5.7. *Let $\tilde{\xi}.$ be a (possibly local) smooth stochastic flow on N with reproducing kernel $\tilde{k}$ and drift $\tilde{A}$. Let $\xi.$, k, and A be corresponding objects on M. Then $\tilde{\xi}.$ lies over $\xi.$, or lies over a flow with the same law as $\xi.$, if and only if $\tilde{k}$ and k and also $\tilde{A}$ and A are p-related.*

Proof. Two flows with the same laws have the same k and $\mathcal{H}$. Intertwining of the flows implies that this is true of their two-point motions and so of the symbols of the generators of these motions, as in Lemma 2.1.1. Since those symbols determine the reproducing kernels by equation (9.15) above, it follows that the reproducing kernels are p-related. By looking at the generators of the one-point motions we see that intertwining also implies the drifts are p-related.

Conversely if the kernels and the drifts are p-related let $p_* : \tilde{\mathcal{H}} \to \mathcal{H}$ be the projection of the reproducing kernel Hilbert spaces induced by p as given by Theorem 9.5.2. If $\{\tilde{W}_t\}_t$ is the Wiener process such that $\tilde{\xi}.$ is the solution flow of the SDE

$$du_t = \tilde{\rho}_{u_t} \circ d\tilde{W}_t + \tilde{A} \, dt,$$

then the p-related SDE

$$dx_t = \rho_{x_t} p_* \circ d\tilde{W}_t + A \, dt$$

has flow with the same law as $\xi.$, since $p_* W.$ is a Wiener process with law given by $\mathcal{H}$. But this flow is intertwined with $\tilde{\xi}.$. $\square$

To simplify the exposition we will consider intertwined stochastic differential equations

$$dy_t = \tilde{X}(y_t) \circ dB_t + \tilde{A}(y_t) \, dt, \tag{9.21}$$
$$dx_t = X(x_t) \circ dB_t + A(x_t) \, dt \tag{9.22}$$

so that $\tilde{X} : N \times \mathbf{R}^m \to TN$ and $X : M \times \mathbf{R}^m \to TM$ are p-related as are the vector fields $\tilde{A}$ and A. By Proposition 9.5.7 above this will allow us to cover the case of intertwined stochastic flows: we can easily extend what follows to allow for infinite dimensional noise.

We shall suppose that $\tilde{X}$ is non-degenerate and determines the given Riemannian structure on N, as then will X for M. Recall from Section 3.3 that the connection $\check{\nabla}$ determined by such an X, the LW connection of the SDE in the terminology of [36], is just the projection of the trivial connection on the product bundle $M \times \mathbf{R}^n$ onto TM. The covariant derivative of a vector field V is given by

$$\check{\nabla}V_u = X(x)d[z \mapsto Y(z)(V(z))](u) \qquad u \in T_xM,$$

where Y is the $\mathbf{R}^m$-valued one-form defined by $Y_x = X(x)^* = [X(x)|_{\ker X(x)^\perp}]^{-1}$. This can be written in terms of the version $k^\#$ of the reproducing kernel used in Chapter 8 as

$$\check{\nabla}V_u = d[z \mapsto k^\#(x, z)(V(z))](u) \qquad u \in T_xM, \tag{9.23}$$

remembering that $k^\#(x, y) = X(y)Y_x$. In particular two different SDE's may have the same reproducing kernel but then their LW connections will be the same. See Theorem 8.1.3 and Remark 8.2.2, or for more details of what follows [36].

This connection has the defining condition

$$\check{\nabla}_u(X(-)(e)) = 0 \qquad u \in T_xM. \tag{9.24}$$

Moreover it is the Levi-Civita connection ∇^M if and only if the exterior derivative, written $d^1Y : \wedge^2 TM \to \mathbf{R}^m$, of Y treated as an $\mathbf{R}^m$-valued one-form satisfies

$$X(x)\,(d^1Y)_x = 0 \qquad x \in M. \tag{9.25}$$

An X determining the Levi-Civita connection can be obtained via Nash's embedding theorem. In fact any metric connection can be obtained from some X by a theorem of Narasimhan & Ramanan [80] with [36], or see [89].

By a *trivial extension* of any $X : M \times \mathbf{R}^m \to TM$, as above, we mean some $X' : M \times \mathbf{R}^{m+q} \to TM$ for some q, again linear on the fibres and smooth, together with a fixed orthogonal projection π of $\mathbf{R}^{m+q}$ onto $\mathbf{R}^m$, or a closed subspace of $\mathbf{R}^m$, such that $X'(x) = X(x) \circ \pi$. The induced connection does not change after taking a trivial extension since this does not change the reproducing kernel k . Consider the SDE

$$dy_t = \tilde{X}(y_t) \circ\ dW_t + \tilde{A}(y_t)\, dt, \tag{9.26}$$
$$dx_t = X(x_t) \circ\ dB_t + A(x_t)\, dt, \tag{9.27}$$

where $W.$ is now a Brownian motion on $\mathbf{R}^{m+q}$ and $B.$ is one on R^m with $\tilde{X}(y) : \mathbf{R}^{m+q} \to T_yN$ and $X(x) : \mathbf{R}^m \to T_xM$. We shall say these are *weakly p-related* if the $\tilde{X}$ is p-related to a trivial extension of X and the vector fields $\tilde{A}$ and A are p-related. Because of Proposition 9.5.7 this is essentially equivalent to their flows being intertwined:

Proposition 9.5.8. *The SDE above are weakly p-related if and only if their repro-ducing kernels $\tilde{k}$ and k are p-related in the sense of equation (9.18).*

Proof. Recall that the reproducing kernels are related to the SDE by $k(x,y) = X(y)X(x)^* : T_x^* M \to T_y M$ and similarly for $\tilde{k}$.

If the SDE's are p-related their reproducing kernels are p-related, and so the same holds if the SDE's are weakly p-related.

For the converse assume that the kernels are p-related. Let $\tilde{\mathcal{H}}$ and $\mathcal{H}$ denote their reproducing kernel Hilbert spaces. By Theorem 9.5.2 the canonical SDE's $\tilde{\rho}. : \tilde{\mathcal{H}} \to TN$ and $\rho. : \mathcal{H} \to TN$ are weakly p-related (extending our definition to cover the infinite dimensional case and using p_*). Let $\tilde{\mathcal{X}} : \mathbf{R}^{m+q} \to \tilde{\mathcal{H}}$ be defined by $e \mapsto \tilde{x}(\cdot)(e)$ with $\mathcal{X} : \mathbf{R}^m \to \mathcal{H}$ defined similarly. For $u \in N$ and $x = p(u)$ we have the commutative diagram:

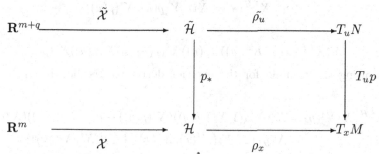

Define $\pi : \mathbf{R}^{m+q} \to \mathbf{R}^m$ by $\pi = \mathcal{X}^* \, p_* \, \tilde{\mathcal{X}}$. This is an orthonormal projection onto a subspace of $\mathbf{R}^m$ because $\tilde{\mathcal{X}}$, $\mathcal{X}$ and p_* are all surjective orthogonal projections. Moreover

$$X(x)\pi(e) = X(x)\mathcal{X}^* p_* \tilde{\mathcal{X}}(e) = \rho_x \mathcal{X}\mathcal{X}^* \, p_* \, \tilde{\mathcal{X}}(e)$$
$$= \rho_x \, p_* \, \tilde{\mathcal{X}}(e) = T_u p \tilde{\rho}_u \tilde{\mathcal{X}}(e)$$
$$= T_u p \tilde{X}(u)(e)$$

as required. □

Theorem 9.5.9. *Consider a Riemannian submersion $p : N \to M$ with a smooth $X : M \times \mathbf{R}^m \to TM$ as above, inducing the Riemannian metric on M. Then there exists $\tilde{X}$ on N which is weakly p-related to X and whose induced connection $\check{\tilde{\nabla}}$ is the Levi-Civita connection if and only if $\check{\nabla}$, the connection induced by X, is the Levi-Civita connection of M, the submersion has totally geodesic fibres, and the horizontal subspace of $p : N \to M$ is integrable.*

Proof. First suppose there exists such an $\tilde{X}$. We can replace X by a trivial exten-sion if necessary and so assume that X and $\tilde{X}$ are p-related. At each $y \in N$ we can decompose $\mathbf{R}^m$ into

$$R^m = K_{p(y)}^{\perp}[\mathbf{R}^m] \oplus Q_y[\mathbf{R}^m] \oplus \tilde{K}_y[\mathbf{R}^m]$$

where, as usual $K_x^{\perp}$ is the orthogonal projection of $\mathbf{R}^m$ onto the orthogonal complement of the kernel of $X(x)$, while $\tilde{K}_y$ is the projection onto the kernel of $\tilde{X}(y)$, and $Q_y = I - K_{p(y)}^{\perp} - \tilde{K}_y$. Fixing y and setting $p(y) = x$, take an orthonormal base $\{e_1, \ldots, e_m\}$ for $\mathbf{R}^m$ with $e_1, \ldots, e_n$ in $K_x^{\perp}[\mathbf{R}^m]$, and $e_{n+1}, \ldots, e_{n+p}$ in the image of Q_y. Here $n = \dim M$ and $n + p = \dim N$. Set $X^j = X(-)e_j$ and let $X^{j,H}$ denote its horizontal lift.

We have, for any i, j,

$$[X^{i,H}, X^{j,H}](y) = \nabla^N_{X^{i,H}(y)} X^{j,H} - \nabla^N_{X^{j,H}(y)} X^{i,H}$$
$$= \tilde{X}(y) \left(d(\tilde{Y} X^{j,H})(y)(X^{i,H}) - d(\tilde{Y} X^{i,H})(y)(X^{j,H}) \right).$$

Now by formula (2.22), for $u \in N$:

$$X^{i,H}(u) = \tilde{X}(u) Y(p(u)) X^i(p(u))$$

giving

$$\tilde{Y}_u X^{i,H}(u) = \tilde{K}^{\perp}(u) Y(p(u)) X^i(p(u)) = Y(p(u)) X^i(p(u)).$$

Thus, using the formula for the exterior derivative recalled in equation (9.29) below:

$$[X^{i,H}, X^{j,H}](y) = \tilde{X}(y) \left(d(Y X^j)(p(y))(X^i(p(y)) - d(Y X^i)(p(y))(X^j(p(y))) \right)$$
$$= \tilde{X}(y) \left(2d^1 Y_x(X^i(x), X^j(x)) + Y_x([X^i, X^j](x)) \right)$$
$$= \tilde{X}(y)(2d^1 Y_x(X^i(x), X^j(x))) + h_y([X^i, X^j](x)).$$

Let P_u^H and P_u^V denote the projections of $T_u N$ onto its horizontal and vertical subspaces for any $u \in N$. From [83] we know that $h_y([X^i, X^j](x)) = P^H([X^{i,H}, X^{j,H}](y))$. Consequently we have

$$P_y^V([X^{i,H}, X^{j,H}](y)) = \tilde{X}(y)(2d^1 Y_x(X^i(x), X^j(x))).$$

In particular since X and $\tilde{X}$ are p-related,

$$0 = Tp\tilde{X}(y)(2d^1 Y_x(X^i(x), X^j(x))) = X(x)(2d^1 Y_x(X^i(x), X^j(x))).$$

Thus $X(x)d^1 Y_x = 0$ for all $x \in M$ and so X induces the Levi-Civita connection.

To prove integrability of the horizontal subspace it is convenient to quote O'Neill again. In [83] he defines a tensor A, which we will write as A^O to distinguish it from shape operators, which gives a bilinear skew-symmetric map $(a, b) \mapsto A_a^O b$: $TN \oplus TN \to TN$ such that, if U and V are horizontal vector fields, then

$$A^O_{U(u)} V(u) = P^V \nabla^N_{U(u)} V = \frac{1}{2} P_u^V([U, V](u)).$$

It will therefore suffice to show that $A_u^O v = 0$ for all horizontal tangent vectors u, v at y.

For this take $U = X^{i,H}$ and $V = X^{j,H}$. Then, using equation (9.24) and O'Neill's result that covariant differentiation commutes with horizontal lifts, [83], we see that $\nabla^N_{X^{i,H}(y)} X^{j,H} = h_y\left(\nabla^M_{X^i(x)} X^j\right) = 0$ if $j \leqslant n$. But $X^{j,H}(y) = 0$ if $j > n$. Thus since A^O is a tensor we have $A^O_{X^{i,H}(y)} X^{j,H}(y) = 0$ for all i, j, proving that A^O vanishes on horizontal vectors.

To see that p has totally geodesic fibres first note that, as our U and V vary through all i and j, so $P^V \nabla^N_{U(y)} V$ determines all values of the second fundamental form of the leaf through y of the horizontal foliation. It therefore vanishes and so correspondingly does the shape operator of the leaf, see Section 9.4. This shape operator is given by the horizontal component of the covariant derivative of each vertical vector, Z say, in a horizontal direction. Thus $P^H \nabla^N_U Z = 0$ for all horizontal vector fields U.

If also U is basic, i.e. the horizontal lift of a vector field on M, then as observed by O'Neill the bracket $[Z, U]$ is vertical and so $\nabla^N_Z U$ must be vertical. In particular this holds for $U = X^{j,H}$. Now take $1 \leqslant j \leqslant n$. Then $\nabla^N \tilde{X}^j$ vanishes at Y by equation (9.24). However $X^{j,H} = \tilde{X}^j - P^V \tilde{X}^j$. Therefore $\nabla^N_{Z(y)} X^{j,H} = -\nabla^N_{Z(y)} P^V \tilde{X}^j$ and so if α^V_y denotes the second fundamental form of the fibre $p^{-1}(x)$ at y we have

$$\alpha^V_y(Z(y), P^V \tilde{X}^j(y)) = P^H \nabla^N_{Z(y)} P^V \tilde{X}^j = 0.$$

Further, if $n + 1 \leqslant j \leqslant n + p$ we have $\nabla^N_{Z(y)} \tilde{X}^j = 0$ by equation (9.24), so that $\alpha^V_y(Z(y), P^V \tilde{X}^j(y)) = 0$ since each $\tilde{X}^j$ is vertical for such j. However $\tilde{X}^1(y), \ldots, \tilde{X}^{n+p}(y)$ span $T_y N$, so this implies that $\alpha^V_y = 0$. Thus the fibres are totally geodesic.

For the converse suppose that X determines the Levi-Civita connection of M and that the horizontal distribution is integrable. Let $X^H : N \times \mathbf{R}^m \to TN$ be the horizontal lift of X. Choose some $\mathcal{X} : N \times \mathbf{R}^q \to TN$ which induces the Riemannian metric for N and its Levi-Civita connection. Set $X^V = P^V \mathcal{X}$. This will induce a connection ∇^V on the vertical tangent bundle VTN which restricts to the Levi-Civita connection of the fibres. Moreover if $v \in T_y N$ and B is a vertical vector field we see immediately that

$$P^V \nabla^N_v B = \nabla^V_v B. \tag{9.28}$$

Define $\tilde{X} : N \times \mathbf{R}^m \times \mathbf{R}^q \to TN$ by $\tilde{X}(u)(e, f) = X^H(u)(e) + X^V(u)(f)$. We claim that $\tilde{X}$ induces the Levi-Civita connection. It is clearly p-related to X.

To prove the claim we will show that $\tilde{X} d^1 \tilde{Y} = 0$. For this, first note that

$$\tilde{Y}_y(v) = (p^* Y)_y(v) + Y^V_y(P^V v) \qquad v \in T_y N$$

where $Y^V_y : VT_y N \to \mathbf{R}^q$ is the adjoint of $X^V(y)$. Observe that $d^1 p^* Y = p^* d^1 Y$ and $\tilde{X}(y)(p^* d^1 Y_y) = h_y\left(X(x) d^1 Y(T_y p-, T_y p-)\right) = 0$ because X induces the Levi-Civita connection. Thus

$$\tilde{X} d^1 \tilde{Y} = \tilde{X} d^1(Y^V P^V).$$

Recall that for vector fields A and B on N and a, possibly vector-valued, one-form ϕ such as $Y^V P^V$,

$$2d^1(\phi)(A(y), B(y)) = d\left(\phi(B(-))\right)(A(y)) - d\left(\phi(A(-))\right)(B(y)) - \phi_y([A,B](y)). \tag{9.29}$$

We can conclude by considering three cases:

(i) *A and B vertical.* Then

$$d^1(Y^V P^V)(A(y), B(y)) = d^1(Y^V|_{p^{-1}(x)})(A(y), B(y))$$

and so $\tilde{X}(y)d^1(Y^V P^V)(A(y), B(y)) = 0$ because X^V induces the Levi-Civita connection on each fibre of p.

(ii) *A and B horizontal.* Then

$$d^1(Y^V P^V)(A(y), B(y)) = -(Y_y^V P^V)([A,B](y)) = 0$$

because the horizontal distribution is assumed to be integrable.

(iii) *A horizontal and B vertical.* Then

$$2d^1(Y^V P^V)(A(y), B(y)) = d\left(Y^V(B(-))\right)(A(y)) - Y_y^V P^V([A,B](y))$$

so that

$$2\tilde{X}(y)d^1(Y^V P^V)(A(y), B(y)) = \nabla_{A(y)}^V B - P^V\left(\nabla_{A(y)}^N B - \nabla_{B(y)}^N A\right)$$
$$= P^V \nabla_{B(y)}^N A$$

using equation (9.28). However $P^V \nabla_{B(y)}^N A = A_{A(y)}^{p^{-1}(x)}(B(y))$ for $A^{p^{-1}(x)}$ the shape operator of the fibre of p through y, and this vanishes because the fibres are totally geodesic. $\qquad\square$

Remark 9.5.10. The conditions on the submersion required to have such commuting SDE's are very strong. See the discussion in the Notes on Chapter 7, Section 7.3. In fact for N complete and M simply connected, N is just the product of M with a Riemannian manifold and p is the projection: using this the sufficiency assertion given above becomes trivial. However the direct proof given here seems more illuminating.

Remark 9.5.11. As described briefly in Remark 7.2.4, and proved in [36], for a Riemannian symmetric space M with natural projection $p : K \to M = K/G$, the right-invariant SDE on the group K lies over an SDE on M which induces the Levi-Civita connection. The right-invariant SDE induces the flat right-invariant connection on K, with adjoint the flat left-invariant connection. Theorem 9.5.9 shows that in general there can be no G-invariant SDE on K which induces the Levi-Civita connection on K.

Bibliography

[1] M. Arnaudon and S. Paycha. Factorisation of semi-martingales on principal fibre bundles and the Faddeev-Popov procedure in gauge theories. *Stochastics Stochastics Rep.*, 53(1-2):81–107, 1995.

[2] Alan Bain and Dan Crisan. *Fundamentals of stochastic filtering*, volume 60 of *Stochastic Modelling and Applied Probability*. Springer, New York, 2009.

[3] D. Bakry and Michel Émery. Diffusions hypercontractives. In *Séminaire de probabilités, XIX, 1983/84*, volume 1123 of *Lecture Notes in Math.*, pages 177–206. Springer, Berlin, 1985.

[4] F. Baudoin. Conditioning and initial enlargement of filtration on a Riemannian manifold. *Ann. Probab.*, 32(3A):2286–2303, 2004.

[5] Fabrice Baudoin. *An introduction to the geometry of stochastic flows.* Imperial College Press, London, 2004.

[6] P. Baxendale. Gaussian measures on function spaces. *Amer. J. Math.*, 98(4):891–952, 1976.

[7] P. Baxendale. Brownian motions in the diffeomorphism groups I. *Compositio Math.*, 53:19–50, 1984.

[8] Ya. I. Belopolskaya and Yu. L. Dalecky. *Stochastic equations and differential geometry*, volume 30 of *Mathematics and its Applications (Soviet Series)*. Kluwer Academic Publishers Group, Dordrecht, 1990. Translated from the Russian.

[9] L. Bérard-Bergery and J.-P. Bourguignon. Laplacians and Riemannian submersions with totally geodesic fibres. *Illinois J. Math.*, 26(2):181–200, 1982.

[10] E. Berger, R. Bryant, and P. Griffiths. The Gauss equations and rigidity of isometric embeddings. *Duke Math. J.*, 50(3):803–892, 1983.

[11] Jean-Michel Bismut. *Large deviations and the Malliavin calculus*, volume 45 of *Progress in Mathematics*. Birkhäuser Boston Inc., Boston, MA, 1984.

[12] Hermann Boerner. *Representations of groups. With special consideration for the needs of modern physics*. Translated from the German by P. G.

Murphy in cooperation with J. Mayer-Kalkschmidt and P. Carr. Second English edition. North-Holland Publishing Co., Amsterdam, 1970.

[13] Jean-Michel Bony, Fabrizio Broglia, Ferruccio Colombini, and Ludovico Pernazza. Nonnegative functions as squares or sums of squares. *J. Funct. Anal.*, 232(1):137–147, 2006.

[14] Raymond Brummelhuis. A counterexample to the Fefferman-Phong inequality for systems. *C. R. Acad. Sci. Paris Sér. I Math.*, 310(3):95–98, 1990.

[15] Z. Brzeźniak and K. D. Elworthy. Stochastic differential equations on Banach manifolds. *Methods Funct. Anal. Topology*, 6(1):43–84, 2000.

[16] A. Carverhill. Conditioning a lifted stochastic system in a product space. *Ann. Probab.*, 16(4):1840–1853, 1988.

[17] A. P. Carverhill and K. D. Elworthy. Flows of stochastic dynamical systems: the functional analytic approach. *Z. Wahrsch. Verw. Gebiete*, 65(2):245–267, 1983.

[18] Man Duen Choi and Tsit Yuen Lam. Extremal positive semidefinite forms. *Math. Ann.*, 231(1):1–18, 1977/78.

[19] Yvonne Choquet-Bruhat, Cécile DeWitt-Morette, and Margaret Dillard-Bleick. *Analysis, manifolds and physics*. North-Holland Publishing Co., Amsterdam, second edition, 1982.

[20] Michael Cranston and Yves Le Jan. Asymptotic curvature for stochastic dynamical systems. In *Stochastic dynamics (Bremen, 1997)*, pages 327–338. Springer, New York, 1999.

[21] Dan Crisan, Michael Kouritzin, and Jie Xiong. Nonlinear filtering with signal dependent observation noise. *Electron. J. Probab.*, 14:no. 63, 1863–1883, 2009.

[22] H. L. Cycon, R. G. Froese, W. Kirsch, and B. Simon. *Schrödinger operators with application to quantum mechanics and global geometry*. Texts and Monographs in Physics. Springer-Verlag, Berlin, study edition, 1987.

[23] R. W. R. Darling. Approximating Ito integrals of differential forms and geodesic deviation. *Z. Wahrsch. Verw. Gebiete*, 65(4):563–572, 1984.

[24] M. H. A. Davis and M. P. Spathopoulos. Pathwise nonlinear filtering for nondegenerate diffusions with noise correlation. *SIAM J. Control Optim.*, 25(2):260–278, 1987.

[25] B. K. Driver. A Cameron-Martin type quasi-invariance theorem for Brownian motion on a compact Riemannian manifold. *J. Functional Analysis*, 100:272–377, 1992.

[26] T. E. Duncan. Some filtering results in Riemann manifolds. *Information and Control*, 35(3):182–195, 1977.

[27] A. Eberle. *Uniqueness and non-uniqueness of semigroups generated by singular diffusion operators*, volume 1718 of *Lecture Notes in Mathematics*. Springer-Verlag, Berlin, 1999.

[28] D. G. Ebin and J. Marsden. Groups of diffeomorphisms and the motion of an incompressible fluid. *Ann. Math.*, pages 102–163, 1970.

[29] Robert J. Elliott and Michael Kohlmann. Integration by parts, homogeneous chaos expansions and smooth densities. *Ann. Probab.*, 17(1):194–207, 1989.

[30] K. D. Elworthy. *Stochastic Differential Equations on Manifolds*. LMS Lecture Notes Series 70, Cambridge University Press, 1982.

[31] K. D. Elworthy. Geometric aspects of diffusions on manifolds. In P. L. Hennequin, editor, *Ecole d'Eté de Probabilités de Saint-Flour XV-XVII, 1985-1987. Lecture Notes in Mathematics 1362*, volume 1362, pages 276–425. Springer-Verlag, 1988.

[32] K. D. Elworthy. Stochastic flows on Riemannian manifolds. In *Diffusion processes and related problems in analysis, Vol. II (Charlotte, NC, 1990)*, volume 27 of *Progr. Probab.*, pages 37–72. Birkhäuser Boston, Boston, MA, 1992.

[33] K. D. Elworthy and W. S. Kendall. Factorization of harmonic maps and Brownian motions. In *From local times to global geometry, control and physics (Coventry, 1984/85), Pitman Res. Notes Math. Ser., 150*, pages 75–83. Longman Sci. Tech., Harlow, 1986.

[34] K. D. Elworthy, Y. Le Jan, and Xue-Mei Li. Equivariant diffusions on principal bundles. In *Stochastic analysis and related topics in Kyoto*, volume 41 of *Adv. Stud. Pure Math.*, pages 31–47. Math. Soc. Japan, Tokyo, 2004.

[35] K. D. Elworthy, Y. Le Jan, and Xue-Mei Li. Concerning the geometry of stochastic differential equations and stochastic flows. In *'New Trends in stochastic Analysis', Proc. Taniguchi Symposium, Sept. 1994, Charingworth, ed. K. D. Elworthy and S. Kusuoka, I. Shigekawa*. World Scientific Press, 1996.

[36] K. D. Elworthy, Y. Le Jan, and Xue-Mei Li. *On the geometry of diffusion operators and stochastic flows, Lecture Notes in Mathematics 1720*. Springer, 1999.

[37] K. D. Elworthy and S. Rosenberg. Homotopy and homology vanishing theorems and the stability of stochastic flows. *Geom. Funct. Anal.*, 6(1):51–78, 1996.

[38] K. D. Elworthy and M. Yor. Conditional expectations for derivatives of certain stochastic flows. In J. Azéma, P.A. Meyer, and M. Yor, editors, *Sem. de Prob. XXVII. Lecture Notes in Maths. 1557*, pages 159–172. Springer-Verlag, 1993.

[39] K. David Elworthy. The space of stochastic differential equations. In *Stochastic analysis and applications*, volume 2 of *Abel Symp.*, pages 327–337. Springer, Berlin, 2007.

[40] M. Émery. *Stochastic calculus in manifolds*. Universitext. Springer-Verlag, Berlin, 1989. With an appendix by P.-A. Meyer.

[41] A. Estrade, M. Pontier, and P. Florchinger. Filtrage avec observation discontinue sur une variété. Existence d'une densité régulière. *Stochastics Stochastics Rep.*, 56(1-2):33–51, 1996.

[42] Maria Falcitelli, Stere Ianus, and Anna Maria Pastore. *Riemannian submersions and related topics*. World Scientific Publishing Co. Inc., River Edge, NJ, 2004.

[43] M. Freidlin. *Functional integration and partial differential equations*, volume 109 of *Annals of Mathematics Studies*. Princeton University Press, Princeton, NJ, 1985.

[44] Kenrô Furutani. On a differentiable map commuting with an elliptic pseudo-differential operator. *J. Math. Kyoto Univ.*, 24:197–203, 1984.

[45] Zhong Ge. Betti numbers, characteristic classes and sub-Riemannian geometry. *Illinois J. Math.*, 36(3):372–403, 1992.

[46] Peter B. Gilkey, John V. Leahy, and Jeonghyeong Park. *Spectral geometry, Riemannian submersions, and the Gromov-Lawson conjecture*. Studies in Advanced Mathematics. Chapman & Hall/CRC, Boca Raton, FL, 1999.

[47] Yuri E. Gliklikh. *Ordinary and stochastic differential geometry as a tool for mathematical physics*, volume 374 of *Mathematics and its Applications*. Kluwer Academic Publishers Group, Dordrecht, 1996. With an appendix by the author and T. J. Zastawniak.

[48] S. I. Goldberg and T. Ishihara. Riemannian submersions commuting with the Laplacian. *J. Differential Geom.*, 13(1):139–144, 1978.

[49] Mikhael Gromov. Carnot-Carathéodory spaces seen from within. In *Sub-Riemannian geometry*, volume 144 of *Progr. Math.*, pages 79–323. Birkhäuser, Basel, 1996.

[50] István Gyöngy. Stochastic partial differential equations on manifolds. II. Nonlinear filtering. *Potential Anal.*, 6(1):39–56, 1997.

[51] Robert Hermann. A sufficient condition that a mapping of Riemannian manifolds be a fibre bundle. *Proc. Amer. Math. Soc.*, 11:236–242, 1960.

[52] J. E. Humphreys. *Introduction to Lie algebras and representation theory*, volume 9 of *Graduate Texts in Mathematics*. Springer-Verlag, New York, 1978. Second printing, revised.

[53] N. Ikeda and S. Watanabe. *Stochastic Differential Equations and Diffusion Processes , second edition.* North-Holland, 1989.

[54] M. Joannides and LeGland F. Nonlinear filtering with continuous time perfect observations and noninformative quadratic variation. In *Proceedings of the 36th IEEE Conference on Decision and Control, San Diego, December 1012, 1997,* page pp. 16451650, 1997.

[55] S. Kobayashi and K. Nomizu. *Foundations of differential geometry, Vol. I.* Interscience Publishers, 1969.

[56] S. Kobayashi and K. Nomizu. *Foundations of differential geometry, Vol. II.* Interscience Publishers, 1969.

[57] I. Kolář, P. W. Michor, and J. Slovák. *Natural operations in differential geometry.* Springer-Verlag, Berlin, 1993.

[58] Hiroshi Kunita. Cauchy problem for stochastic partial differential equations arising in nonlinear filtering theory. *Systems Control Lett.,* 1(1):37–41, 1981/82.

[59] Hiroshi Kunita. *Stochastic flows and stochastic differential equations,* volume 24 of *Cambridge Studies in Advanced Mathematics.* Cambridge University Press, Cambridge, 1990.

[60] S. Kusuoka. Degree theorem in certain Wiener Riemannian manifolds. Stochastic analysis, Proc. Jap.-Fr. Semin., Paris/France 1987, Lect. Notes Math. 1322, 93-108 , 1988.

[61] Serge Lang. *Introduction to differentiable manifolds.* Universitext. Springer-Verlag, New York, second edition, 2002.

[62] Joan-Andreu Lázaro-Camí and Juan-Pablo Ortega. Reduction, reconstruction, and skew-product decomposition of symmetric stochastic differential equations. *Stoch. Dyn.,* 9(1):1–46, 2009.

[63] Yves Le Jan and Olivier Raimond. Solutions statistiques fortes des équations différentielles stochastiques. *C. R. Acad. Sci. Paris Sér. I Math.,* 327(10):893–896, 1998.

[64] Yves Le Jan and Olivier Raimond. Integration of Brownian vector fields. *Ann. Probab.,* 30(2):826–873, 2002.

[65] Rémi Léandre. Applications of the Malliavin calculus of Bismut type without probability. *WSEAS Trans. Math.,* 5(11):1205–1210, 2006.

[66] Y. Le Jan and S. Watanabe. Stochastic flows of diffeomorphisms. In *Stochastic analysis (Katata/Kyoto, 1982), North-Holland Math. Library, 32,,* pages 307–332. North-Holland, Amsterdam, 1984.

[67] S. Lemaire. Invariant jets of a smooth dynamical system. *Bull. Soc. Math. France,* 129(3):379–448, 2001.

[68] Xue-Mei Li. Stochastic differential equations on noncompact manifolds: moment stability and its topological consequences. *Probab. Theory Related Fields*, 100(4):417–428, 1994.

[69] M. Liao. Factorization of diffusions on fibre bundles. *Trans. Amer. Math. Soc.*, 311(2):813–827, (1989).

[70] Ming Liao. Decomposition of stochastic flows and Lyapunov exponents. *Probab. Theory Related Fields*, 117(4):589–607, 2000.

[71] Andre Lichnerowicz. Quelques théoremes de géométrie différentielle globale. *Comment. Math. Helv.*, 22:271–301, 1949.

[72] S. V. Lototsky and B. L. Rozovskii. Wiener chaos solutions of linear stochastic evolution equations. *Ann. Probab.*, 34(2):638–662, 2006.

[73] Sergey Lototsky, Remigijus Mikulevicius, and Boris L. Rozovskii. Nonlinear filtering revisited: a spectral approach ii. In *Proc. 35th IEEE Conf. on Decision and Control*, volume 4 of *Kobe,Japan,1996*, pages 4060–4064. Omnipress, Madison,WI., 1996.

[74] Sergey Lototsky and Boris L. Rozovskii. Recursive multiple Wiener integral expansion for nonlinear filtering of diffusion processes. In *Stochastic processes and functional analysis (Riverside, CA, 1994)*, volume 186 of *Lecture Notes in Pure and Appl. Math.*, pages 199–208. Dekker, New York, 1997.

[75] Henry P. McKean. *Stochastic integrals*. AMS Chelsea Publishing, Providence, RI, 2005. Reprint of the 1969 edition, with errata.

[76] P. W. Michor. *Gauge theory for fiber bundles*, volume 19 of *Monographs and Textbooks in Physical Science. Lecture Notes*. Bibliopolis, Naples, 1991.

[77] D Mitter, S.K.and Ocone. Multiple integral expansions for nonlinear filtering. In *18th IEEE Conf. on decision and Control including Symp. on Adaptive Processes*, volume 18 of *1979*, pages 329–334. IEEE, 1979.

[78] S.A. Molchanov. Diffusion processes and riemannian geometry. *Russian Mathematical Surveys*, 30:2–63, 1975.

[79] Richard Montgomery. *A tour of subriemannian geometries, their geodesics and applications*, volume 91 of *Mathematical Surveys and Monographs*. American Mathematical Society, Providence, RI, 2002.

[80] M. S. Narasimhan and S. Ramanan. Existence of universal connections. *American J. Math.*, 83, 1961.

[81] D Ocone. Multiple integral expansions for nonlinear filtering. *Stochastics*, 10(1):1–30, 1983.

[82] O. A. Oleĭnik. On linear equations of the second order with a non-negative characteristic form. *Mat. Sb. (N.S.)*, 69 (111):111–140, 1966.

[83] Barrett O'Neill. The fundamental equations of a submersion. *Michigan Math. J.*, 13:459–469, 1966.

[84] E. Pardoux. Nonlinear filtering, prediction and smoothing. In *Stochastic systems: the mathematics of filtering and identification and applications (Les Arcs, 1980)*, volume 78 of *NATO Adv. Study Inst. Ser. C: Math. Phys. Sci.*, pages 529–557. Reidel, Dordrecht, 1981.

[85] Étienne Pardoux. Filtrage non linéaire et équations aux dérivées partielles stochastiques associées. In *École d'Été de Probabilités de Saint-Flour XIX— 1989*, volume 1464 of *Lecture Notes in Math.*, pages 67–163. Springer, Berlin, 1991.

[86] Alberto Parmeggiani. A class of counterexamples to the Fefferman-Phong inequality for systems. *Comm. Partial Differential Equations*, 29(9-10):1281–1303, 2004.

[87] E. J. Pauwels and L. C. G. Rogers. Skew-product decompositions of Brownian motions. In *Geometry of random motion (Ithaca, N.Y., 1987)*, volume 73 of *Contemp. Math.*, pages 237–262. Amer. Math. Soc., 1988.

[88] Monique Pontier and Jacques Szpirglas. Filtering with observations on a Riemannian symmetric space. In *Stochastic differential systems (Bad Honnef, 1985)*, volume 78 of *Lecture Notes in Control and Inform. Sci.*, pages 316–329. Springer, Berlin, 1986.

[89] D. Quillen. Superconnection character forms and the Cayley transform. *Topology*, 27(2):211–238, 1988.

[90] Michael Reed and Barry Simon. *Methods of modern mathematical physics. I.* Academic Press Inc. [Harcourt Brace Jovanovich Publishers], New York, second edition, 1980. Functional analysis.

[91] Daniel Revuz and Marc Yor. *Continuous martingales and Brownian motion*, volume 293 of *Grundlehren der Mathematischen Wissenschaften [Fundamental Principles of Mathematical Sciences]*. Springer-Verlag, Berlin, third edition, 1999.

[92] L. C. G. Rogers and D. Williams. *Diffusions, Markov processes, and martingales. Vol. 2.* Cambridge Mathematical Library. Cambridge University Press, Cambridge, 2000.

[93] S. Rosenberg. *The Laplacian on a Riemannian manifold*, volume 31 of *London Mathematical Society Student Texts*. Cambridge University Press, Cambridge, 1997. An introduction to analysis on manifolds.

[94] Paulo R. C. Ruffino. Decomposition of stochastic flows and rotation matrix. *Stoch. Dyn.*, 2(1):93–107, 2002.

[95] D. Stroock and S. R. S. Varadhan. On degenerate elliptic-parabolic operators of second order and their associated diffusions. *Comm. Pure Appl. Math.*, 25:651–713, 1972.

[96] D. W. Stroock. *Lectures on stochastic analysis: diffusion theory*, volume 6 of *London Mathematical Society Student Texts*. Cambridge University Press, Cambridge, 1987.

[97] D. W. Stroock and S. R. S. Varadhan. *Multidimensional diffusion processes*, volume 233 of *Grundlehren der Mathematischen Wissenschaften [Fundamental Principles of Mathematical Sciences]*. Springer-Verlag, Berlin, 1979.

[98] H. J. Sussmann. Orbits of families of vector fields and integrability of distributions. *Trans. Amer. Math. Soc.*, 180:171–188, 1973.

[99] Kazuaki Taira. *Diffusion processes and partial differential equations*. Academic Press Inc., Boston, MA, 1988.

[100] Daniel Tataru. On the Fefferman-Phong inequality and related problems. *Comm. Partial Differential Equations*, 27(11-12):2101–2138, 2002.

[101] J. C. Taylor. Skew products, regular conditional probabilities and stochastic differential equation: a technical remark. In *Séminaire de probabilités, XXVI, Lecture Notes in Mathematics 1526, ed. J. Azéma, P. A. Meyer and M. Yor*, pages 113–126, 1992.

[102] B Tsirelson. Filtrations of random processes in the light of classification theory. (i). a topological zero-one law. *arXiv:math.PR/0107121*, 2001.

[103] A. Ju. Veretennikov and N. V. Krylov. Explicit formulae for the solutions of stochastic equations. *Mat. Sb. (N.S.)*, 100(142)(2):266–284, 336, 1976.

[104] Jaak Vilms. Totally geodesic maps. *J. Differential Geometry*, 4:73–79, 1970.

[105] Bill Watson. Manifold maps commuting with the Laplacian. *J. Differential Geometry*, 8:85–94, 1973.

[106] Bill Watson. δ-commuting mappings and Betti numbers. *Tôhoku Math. J. (2)*, 27(2):135–152, 1975.

Index

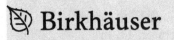
Frontiers in Mathematics

This series is designed to be a repository for up-to-date research results which have been prepared for a wider audience. Graduates and postgraduates as well as scientists will benefit from the latest developments at the research frontiers in mathematics and at the "frontiers" between mathematics and other fields like computer science, physics, biology, economics, finance, etc.

Advisory Board

Leonid Bunimovich (Atlanta), Benoît Perthame (Paris), Laurent Saloff-Coste (Rhodes Hall), Igor Shparlinski (Sydney), Wolfgang Sprössig (Freiberg), Cédric Villani (Lyon)

■ **Østvær, P.A.**, Homotopy Theory of C*-Algebras (2010). ISBN 978-3-0346-0564-9

Homotopy theory and C*-algebras are central topics in contemporary mathematics. This book introduces a modern homotopy theory for C*-algebras.
One basic idea of the setup is to merge C*-algebras and spaces studied in algebraic topology into one category comprising C*-spaces. These objects are suitable fodder for standard homotopy theoretic moves, leading to unstable and stable model structures. With the foundations in place one is led to natural definitions of invariants for C*-spaces such as homology and cohomology theories, K-theory and zeta-functions. The text is largely self-contained. It serves a wide audience of graduate students and researchers interested in C*-algebras, homotopy theory and applications.

■ **Borsuk, M.**, Transmission Problems for Elliptic Second-Order Equations in Non-Smooth Domains (2010).
ISBN 978-3-0346-0476-5

The goal of this book is to investigate the behavior of weak solutions of the elliptic transmission problem in a neighborhood of boundary singularities: angular and conic points or edges. This problem is discussed for both linear and quasilinear equations. A principal new feature of this book is the consideration of our estimates of weak solutions of the transmission problem for linear elliptic equations with minimal smooth coeciffients in n-dimensional conic domains. Only few works are devoted to the transmission problem for quasilinear elliptic equations.

Therefore, we investigate the weak solutions for general divergence quasilinear elliptic second-order equations in n-dimensional conic domains or in domains with edges.
The basis of the present work is the method of integro-differential inequalities. Such inequalities with exact estimating constants allow us to establish possible or best possible estimates of solutions to boundary value problems for elliptic equations near singularities on the boundary. A new Friedrichs–Wirtinger type inequality is proved and applied to the investigation of the behavior of weak solutions of the transmission problem.
All results are given with complete proofs. The book will be of interest to graduate students and specialists in elliptic boundary value problems and applications.

■ **Duggal, K.L. / Sahin, B.**, Differential Geometry of Lightlike Submanifolds (2010).
ISBN 978-3-0346-0250-1

This is the first systematic account of the main results in the theory of lightlike submanifolds of semi-Riemannian manifolds which have a geometric structure, such as almost Hermitian, almost contact metric or quaternion Kähler. Using these structures, the book presents interesting classes of submanifolds whose geometry is very rich.
The book also includes hypersurfaces of semi-Riemannian manifolds, their use in general relativity and Osserman geometry, half-lightlike submanifolds of semi-Riemannian manifolds, lightlike submersions, screen conformal submersions, and their applications in harmonic maps.